Network Optimization: An Introduction to the Network Reconstruction Approach

Network Reconstruction Approach to Optimization

RIVER PUBLISHERS SERIES IN COMMUNICATIONS AND NETWORKING

The "River Publishers Series in Communications and Networking" is a series of comprehensive academic and professional books which focus on communication and network systems. Topics range from the theory and use of systems involving all terminals, computers, and information processors to wired and wireless networks and network layouts, protocols, architectures, and implementations. Also covered are developments stemming from new market demands in systems, products, and technologies such as personal communications services, multimedia systems, enterprise networks, and optical communications.

The series includes research monographs, edited volumes, handbooks and textbooks, providing professionals, researchers, educators, and advanced students in the field with an invaluable insight into the latest research and developments.

Topics included in this series include:-

- Communication theory
- Multimedia systems
- Network architecture
- Optical communications
- Personal communication services
- Telecoms networks
- Wifi network protocols

For a list of other books in this series, visit www.riverpublishers.com

Network Optimization: An Introduction to the Network Reconstruction Approach

Network Reconstruction Approach to Optimization

Elias Munapo

North West University, South Africa

Santosh Kumar

RMIT University, Australia

Philimon Nyamugure

NUST, Bulawayo, Zimbabwe

Trust Tawanda

NUST, Bulawayo, Zimbabwe

River Publishers

Routledge
Taylor & Francis Group
NEW YORK AND LONDON

Published 2025 by River Publishers
River Publishers
Broagervej 10, 9260 Gistrup, Denmark
www.riverpublishers.com

Distributed exclusively by Routledge
605 Third Avenue, New York, NY 10017, USA
4 Park Square, Milton Park, Abingdon, Oxon OX14 4RN

Network Optimization: An Introduction to the Network Reconstruction Approach / by Elias Munapo, Santosh Kumar, Philimon Nyamugure, Trust Tawanda.

Routledge is an imprint of the Taylor & Francis Group, an informa business

ISBN 978-87-7004-742-5 (hardback)
ISBN 978-87-4380-076-7 (paperback)
ISBN 978-87-7004-744-9 (online)
ISBN 978-87-7004-743-2 (master ebook)

Dedication

To our students (past, present, and future) and to our families.
EM, SK, PN, TT

Contents

Preface

This book is a collection of some optimization studies carried out by the authors in network theory and its applications. A common thread joining these various studies is a reconstruction concept that has been applied differently to suit different situations discussed in this book. Each chapter presents a different situation. The idea of reconstruction varies, depends on the goal to be attained and a suitable feasible reconstruction, utilizing the mathematical structure of the situation under study. Reconstruction is an art that utilizes mathematical properties associated with the given structure and the objective we have. Here we illustrate two simple and well-known situations, which are well documented. However, our purpose is to relook at them with an eye of reconstruction.

1. Consider a *Gaussian elimination approach* to solve a system of simultaneous equations. For example, for a system of eqn (P.1), the object is to find the value of three variables, $x_1,\ x_2,$ and x_3 from the three given equations.

$$\begin{aligned} -\,x_2 + x_3 &= 3 \\ x_1 - x_2 - x_3 &= 0 \\ -\,x_1 - x_3 &= -3. \end{aligned} \tag{P.1}$$

The system (P.1) can be rewritten in matrix form as given in eqn (P.2).

$$\begin{bmatrix} 0 & -1 & 1 \\ 1 & -1 & -1 \\ -1 & 0 & -1 \end{bmatrix} \begin{bmatrix} x_1 \\ x_2 \\ x_3 \end{bmatrix} = \begin{bmatrix} 3 \\ 0 \\ -3 \end{bmatrix}. \tag{P.2}$$

By simple elementary processes, the system (P.2) can be rewritten in an equivalent form as given by eqn (P.3)

$$\begin{bmatrix} 1 & -1 & -1 \\ 0 & -1 & 1 \\ 0 & 0 & -3 \end{bmatrix} \begin{bmatrix} x_1 \\ x_2 \\ x_3 \end{bmatrix} = \begin{bmatrix} 0 \\ 3 \\ -6 \end{bmatrix}. \tag{P.3}$$

Since the system (P.3) is equivalent to eqn (P.2) or (P.1), the solution of eqn (P.3) will be the same as that of eqn (P.1) or (P.2). However, (P.3) system being in an echelon form, it can be easily solved by back substitution leading to:

$$x_1 = 1,\ x_2 = -1,\ x_3 = 2. \tag{P.4}$$

Here the requirement was to find the solution to the system of equations (P.1), which became obvious when the same system is rewritten or reconstructed in the form (P.3). This was a simple illustration of reconstruction, and it is known as Gaussian elimination process.

The above simple illustration indicates that the idea of reconstruction is not new; it can be traced easily in the literature in many different forms. Gaussian elimination has been discussed in many books.

The method is named after Carl Friedrich Gauss (1777–1855), picture on the right. It is a process of row reduction on a matrix by using a sequence of elementary row operations to modify the matrix until it is transformed in a desirable echelon form and the solution of the unknowns becomes obvious.

Let us consider another situation from the field of mathematical programming.

2. Consider **a** *linear program in Operations Research* given by eqn (P.5), (P.6), and (P.7).

$$\text{Max } z = f(x) = \sum_{j=1}^{n} c_j x_j \tag{P.5}$$

$$\text{subject to: } \sum_{j=1}^{n} a_{ij} x_j \leq b_i,\ i = 1,\ 2,\ \ldots,\ m \tag{P.6}$$

$$x_j \geq 0,\ \forall j,\ j = 1, 2, \ldots, n. \tag{P.7}$$

The above linear programming model is versatile, and has abstraction that keeps on taking different forms to represent different industrial situations. This LP model can be solved by the well-known simplex method by Dantzig; see Nash (2000). This method involves writing the given set of constraints in different forms, until we have identified the optimal solution.

Here is a simple numerical illustration.

$$\begin{aligned} &\text{Max } z = 5x_1 + 4x_2 \\ \text{Subject to: } &3x_1 + 2x_2 \leq 3 \\ &4x_1 + 3x_2 \leq 5 \\ &x_1,\ x_2 \geq 0. \end{aligned} \tag{P.8}$$

After introduction of two slack variables x_3 and x_4 in the above two constraints, the problem (P.8) can be arranged as shown in Table P1.

Table P1 Simplex Tableau 1 for the problem (P8).

C_j	5	4	0	0	RHS
	x_1	x_2	x_3	x_4	
x_3	3	2	1	0	3
x_4	4	3	0	1	5
$Z_j - C_j$	−5	−4	0	0	0

Table P1 will give rise to Table P2.

Table P2 Upgraded simplex Tableau 2.

C_j	5	4	0	0	RHS
	x_1	x_2	x_3	x_4	
x_1	1	2/3	1/3	0	1
x_4	0	1/3	−4/3	1	1
$Z_j - C_j$	0	−2/3	5/3	0	5

Finally from Table P2, we get Table P3.

This leads to the optimal solution. However, all this is nothing but reconstruction of the same given set of constraints. In Table P1, we expressed the given set of constraints in the form of variables x_3 and x_4 with the value of the objective function equal to zero. In Table P2, that same sets of constraints were expressed in the form of variables x_1 and x_4 resulting in objective function value equal to 5 and finally in Table P3, we expressed the same

Table P3 The final simplex tableau.

C_j	5	4	0	0	RHS
	x_1	x_2	x_3	x_4	
x_2	3/2	1	1/2	0	3/2
x_4	−1/2	0	−3/2	1	1/2
$Z_j - C_j$	1	0	5/3	0	6

set of constraints as a function of variables, x_2 and x_4. This resulted in the objective function value equal to 6 and we concluded that value as the optimal solution.

This book is our humble attempt for analysis of network-related optimization problems using the reconstruction concept in networks. All these approaches were developed by the authors.

We have discussed recent developments in solution approaches for an assignment and transportation models, network routing problems, the well-known spanning tree, and the travelling salesman problem. We have left for the reader a question for them to decide: Is the travelling salesman problem really NP hard? We have considered a few reliability-related problems in stochastic networks, and discussed the TANYAKUMU approach for some network touting problems. Network reliability and flows in network have been studied by using network reconstruction.

The idea of reconstruction is an interesting mathematical concept that deserves more attention, and we hope readers will do justice and investigate further potential of this useful mathematical concept.

This book is our humble attempt for analysis of network-related optimization problems using reconstruction approaches developed by the authors for network optimization. The book is composed of the following eight chapters:

Chapter 1: Network reconstruction – Unification of an assignment and transportation models and accelerating the search process.

Chapter 2: Reconstruction approach for the shortest and the $K\text{th}$ shortest routes in networks.

Chapter 3: Network reconstruction approach to minimum spanning tree under the index restriction.

Chapter 4: Path through '*k*' specified nodes or '*k*' specified links in networks – A reconstruction approach.

Chapter 5: Is determination of the travelling salesman tour an NP hard problem? Some polynomial time reconstruction approaches.

Chapter 6: TANYAKUMU Labelling method – Exact algorithm for the travelling salesman problem (TSP) and equality generalized travelling salesman problem (E-GTSP).

Chapter 7: Some reliability routing problems and their evaluation by network reconstruction.

Chapter 8: Reconstruction approach to reliability and maximum flows in networks

Elias Munapo, Santosh Kumar, Philimon Nyamugure, and Trust Tawanda

https://en.wikipedia.org> wiki > Simplex_algorithm (downloaded 30 Dec 2023)
http://Wikipedia.org/wiki/Carl_Friedrich_Gauss (downloaded 30 Dec 2023)

John C. Nash (2000) The (Dantzig) Simplex Method for Linear Programming, The top 10 algorithms, Computing in Science and Engineering, Jan/Feb pp 29–31.

Acknowledgement

This book is based on recent research contributions by the authors and their research associates. This work was reported in the form of research papers in various international journals, book chapters, and conference presentations. We express our gratitude to all those numerous referees for their constructive suggestions, searching comments, and critical questions that helped us to improve our understanding in this fascinating field of networks optimization and their applications. Their queries were catalysts for our better understanding, and we thank them all from the bottom of our heats. Our students and colleagues from different institutions have very special places in our lives as they were helpful in redirecting us to the right path, when we found ourselves derailed.

We would like to express our gratitude to our respective universities, i.e., the North-West University, South Africa, RMIT University, Australia, and the National University of Science and Technology, Zimbabwe for their support and encouragement.

We express our gratitude to Professor Mangey Ram for his support and encouragement.

We express our gratitude to all our family members for their patience and understanding.

Elias Munapo

Santosh Kumar

Philimon Nyamugure

Trust Tawanda

List of Figures

List of Tables

List of Abbreviations

E-GTSP	Equality generalized travelling salesman problem
LP	Linear programming
MST	Minimum spanning tree
PMST	Partially minimum spanning tree
SP	Shortest path
SPP	Shortest path problems
TPR	Terminal pair reliability
TSP	Travelling salesman problem
WDN	Water distribution network

1

Network Reconstruction – Unification of Assignment and Transportation Models and Accelerating the Search

Abstract

This chapter considers two special linear programming (LP) models, which are not suitable for solution by the conventional simplex method of LP, as both are degenerate LP problems. These two models, known as assignment and transportation models can be easily represented by a network and can be seen as a network optimization problem. Since the LP approach was not suitable, many special efficient methods were developed for solving these models. Among them 'the Hungarian method of assignment' is well known for analysis of the assignment model and a dual LP-based method was developed for the transportation model. The Hungarian method of assignment has been revisited as a network reconstruction approach and it has been extended for solving the transportation model, thus a unification of these two models has been achieved. Special approaches for these two models are well documented in OR Literature, which are not presented here. However, in this chapter, recent developments in this area have been discussed, including the recent accelerating search has been explained and illustrated.

Keywords: Optimization by network reconstruction, link-weight reconstruction in networks, assignment and transportation models, Hungarian method of assignment, modified Hungarian method for a transportation model, search acceleration.

Notations

1. C_{ij} denotes the cost of assignment of job i to the technician j, where $i = 1, 2, \ldots, n$ and $j = 1, 2, \ldots, n$ for the assignment model and it also

represents the cost of transportation per unit from supply point i to the destination *j*.

2. a_i denotes the number of units available at the supply point $i = 1, 2, \ldots, m$ in the transportation model and in the assignment model $a_i = 1$ for all i.
3. b_j denotes the number of units required at the destination $j = 1, 2, \ldots, n$ in the transportation model and in the assignment model $b_j = 1$, for all j, where $m = n$.
4. C_{ij} denotes cost of assigning technician i to job j, where $i, j = 1, 2, \ldots, m$ and $m = n$ in the assignment model and for the transportation model it denotes transportation cost per unit from supply point i to the demand point j, here $i = 1, 2, \ldots, m$ and $j = 1,\ 2,\ \ldots,\ n$.
5. x_{ij} denotes the decision variables $\forall(i, j)$.

1.1 Introduction

In this chapter, we use the reconstruction idea and point out that reconstruction may happen in different forms in different situations. The idea behind a reconstruction is to make a given situation relatively in some easy form such that it can unfold the answer hidden behind those mathematical statements. In the context of a classical assignment model, some ideas that were developed by two Hungarian Mathematicians were used by Kuhn (1955, 2005) for solving an assignment model in operations research. We reclassify Kuhn's approach as a reconstruction approach and extend Kuhn's approach to solve the transportation model. Both these models and their solution approaches have been discussed in OR books, for example, the classical approaches have been discussed by Taha (2017), Hillier and Lieberman (2015) and Winston (2004). A new approach, such as the 'The Method of Subtractions' for the transportation model and the 'See-Saw' approach for the assignment model has been discussed by Munapo and Kumar (2022). In this chapter, these two special OR models have been unified unlike the earlier treatment, where they were considered as two separate models, and their solutions were achieved by using different techniques.

The assignment and transportation models are well-known, and they have many industrial applications in production planning, telecommunication, scheduling, military operations etc. Since both models are essentially degenerate linear programs, better approaches were developed for solving these two models. These special methods for the assignment and transportation models are well documented in many OR books. Mathematical models for both problems are totally unimodular, hence both models have an integer optimal solution. Method for the assignment model was proposed by Kuhn (1955) and he called his approach, 'The Hungarian method for assignment.' His approach requires no proof of optimality; hence the approach is free of difficulties associated with a degenerate LP solution. However, for the transportation model, a special method was developed that used the properties of the dual of the transportation LP model and developed an approach that requires an initial feasible solution, a testing procedure for optimality and if the solution does not satisfy optimality conditions, a process to move to a better solution. This process is repeated until the optimal solution is reached. The efficiency of this approach is dependent on the initial starting solution and the number of basic variables, which is required to be $(m + n - 1)$, where m denotes the number of rows and n denotes the number of columns in the transportation model. Therefore, many methods were developed to find an initial good solution, which was subjected to an optimality test and if the solution failed the test for optimality, iteratively the solution was improved until it satisfied the optimality test. For a degenerate solution, computational load for the test of optimality increases and makes it more difficult to establish the optimality of such a degenerate solution. Degeneracy in LP or for this special class of model can cause difficulties as the value of the objective function may not improve in successive iterations. In many instances, one may have reached the optimal solution, but the test of optimality is unable to recognize its optimality due to degeneracy. The assignment and transportation are both degenerate with respect to a general LP model. The order of degeneracy for the transportation problem as an LP model is 1 or more and for the assignment model, degeneracy with respect to a LP is of the order 'n'.

The assignment model is a special class of linear programming model, for which many solution approaches have been put forward by many researchers, for example, see Aboli et al. (2020), Zhang et al. (2019), Kumar, Ncube and Munapo (2003), Kuhn (1955, 2005), Tomizawa (1971).

In this chapter, an assignment model is briefly considered and without any loss of generality, a balanced transportation problem is seen as an assignment model, and the Hungarian method of assignment is modified to solve the transportation model. Although the unification of transportation and assignment was established earlier by Munapo et al. (2012), however, this chapter establishes a total unification among these two OR models. For this chapter, the approach used here is classified as a reconstruction approach. Recently, Munapo (2020) has introduced some acceleration ideas for determination of the optimal solution of the assignment model, which have been discussed and applied to solve not only the assignment model, but these ideas have been extended to solve the transportation model also. Each case is illustrated by a numerical example.

1.2 Assignment Models

An assignment model in OR can be represented by a network diagram shown in Figure 1.1.

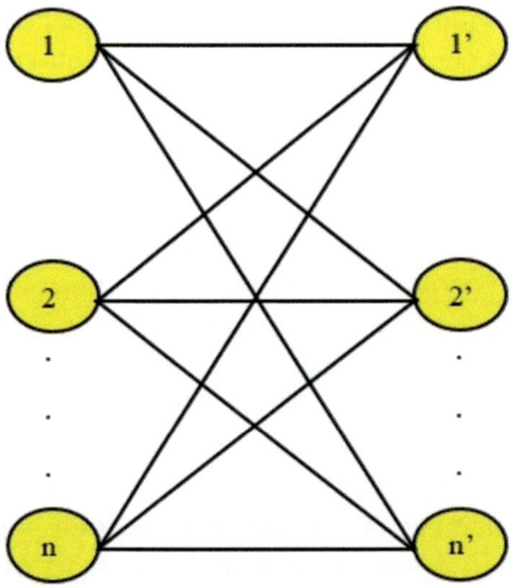

Figure 1.1 *n*-person, *n*'-job assignment network.

Here 1, 2, …, *n* represents a set of persons and 1', 2', …, *n*' represents another set of jobs. The cost of assigning job *j*' to person *j* is represented by C_{jj}, where j and $j'$are $= 1,\ 2, \ldots, n$.

The above (nXn) assignment problem can be expressed by the following mathematical model:

$$\text{Min } Z = \sum_{i=1}^{n}\sum_{j=1}^{n} C_{ij}x_{ij}$$

$$\sum_{j=1}^{n} x_{ij} = 1,\ \sum_{i=1}^{n} x_{ij} = 1,\ i = 1,\ 2, \ldots, n; j = 1,\ 2, \ldots, n.$$

$$x_{ij} = 0 \text{ or } 1 \text{ for all } i \text{ and } j. \tag{1.1}$$

The model (1.1) represents 'n' jobs that are to be completed by an equal number of technicians such that each job is assigned to a technician and each technician must get one job, and the total cost of completing all jobs is minimized. Note C_{ij} denotes the cost of assignment of job i to the technician j, where $i = 1,\ 2, \ldots, n$ and $j = 1, 2, \ldots, n$.

The cost matrix in the model (1.1) can be shown by eqn (1.2).

$$\begin{bmatrix} C_{11} & \ldots & C_{1n} \\ \ldots & \ldots & \ldots \\ C_{n1} & \ldots & C_{nn} \end{bmatrix}. \tag{1.2}$$

The key idea behind the solution approach developed by Kuhn (1955) is based on the following concepts:

1.2.1 Concept

If any constant is added or subtracted from each element in a particular row or column of the assignment cost matrix, relative merit of various technicians with respect to a particular job will remain unchanged and similarly relative merit of various jobs with respect to a particular technician will remain unaffected. This additional number added or subtracted is a reconstruction of various costs.

Concept 1.2.1 is applied to the assignment model by subtracting the minimum in each row from all other elements in that row, thus creating a zero in each row, while relative merits are maintained. The same subtracting process is applied to each column. Thus, each row and column has at least one zero. If in these zero positions, one gets a feasible solution confined to them, that solution becomes an optimal solution with respect to the transformed cost matrix, zero cost is a natural minimum. If these zero elements do not give rise to a feasible solution, some more zero elements must be created. This leads to concept 1.2.2.

1.2.2 Concept

Reconstructed cost elements are maintained at non-negative level, so that zero becomes a natural minimum. Therefore, if an assignment in the transformed cost matrix is such that the total cost is zero, it will be a natural minimum in the transformed cost matrix and the corresponding assignment in the original cost matrix will lead to the minimum total cost solution, as was required.

Each time an assignment is made, a line is drawn to indicate that no further assignment is possible in that row or a column. This leads to concept 1.2.3.

1.2.3 Concept

If a minimum positive element is subtracted from each row which has not been crossed out, and the same minimum is added to each element in a crossed-out column, once again relative merits are maintained, however at the same time all elements remain non-negative and at least one more independent zero position is created.

These three concepts have been organized by Kuhn (1955) as the Hungarian method of assignment, which is well documented in the OR books. The steps of the algorithm are:

Step 1: Subtract the smallest element in each row from all other elements in that row and create a zero entry in each row. Similarly, once again, subtract the smallest element in each column from all other elements in that column. This will create at least one zero in each column.

Step 2: Draw lines through rows and columns and cover all 0 entries by the minimum number of lines.

Step 3: If the number of lines is equal to '*n*,' the order of the given assignment problem, the search is over by making appropriate assignments at 0 entries. If the number of lines is less than '*n*', find the smallest positive entry > 0, and do the following:

3.1 Subtract this value from each element which has not been crossed out.

3.2 Keep the elements as they are on a single line.

3.3 Add this minimum value to an element that is on two lines.

This operation will add at least one more independent zero and return to Step 2.

The above steps were developed by Kuhn (1955) and are well documented in the OR literature, see Taha (2017) and Winston (2004).

Munapo and Kumar (2022) developed a 'See-Saw' approach for the assignment model and a method of subtractions for obtaining a feasible good starting solution for the transportation model. In this chapter, we relabel the Hungarian method of assignment as a reconstruction approach and extend the reconstruction idea of assignment to the transportation model in OR. Further, we discuss some more recent ideas to accelerate the search process, which were developed by Munapo (2020).

1.3 The Transportation Models

The transportation model can be represented by a network diagram as shown in Figure 1.2.

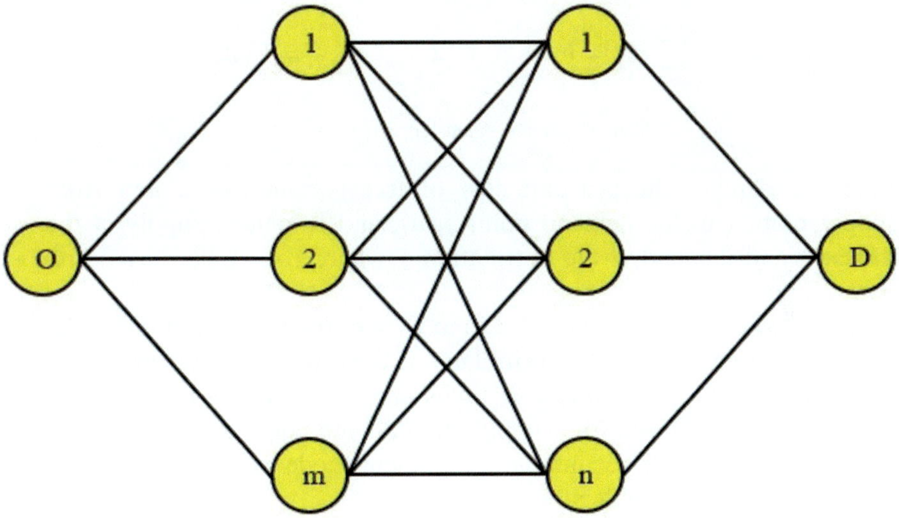

Figure 1.2 Network presentation of a transportation model.

In Figure 1.2, i represents a supply point or a warehouse, where $i = 1, 2, \ldots, m$.

The demand point is represented by j, where $j = 1, 2, \ldots, n$. The links (Oi), $i = 1, 2, \ldots, m$ represent links with a capacity through which a_i number of units can flow and similarly (jD), $j = 1, 2, \ldots, n$ represent links with capacity b_j. The total supply at 'O' and the demand at 'D' is given by M.

A transportation model is like the assignment model, where supplies and demands are not one unit but is some known positive integer quantity ≥ 1, denoted by a_i, b_j, $i = 1, 2, \ldots, m$ and $j = 1, 2, \ldots, n$. And for the balanced transportation models we have

$$\sum_{i=1}^{m} a_i = \sum_{j=1}^{n} b_j = M. \tag{1.3}$$

The quantity M is a known integer number.

A mathematical formulation of the mxn transportation model is given by:

$$\text{Min } Z = \sum_{i=1}^{m}\sum_{j=1}^{n} C_{ij}x_{ij}$$

$$\sum_{j=1}^{n} x_{ij} = a_i,\ \sum_{i=1}^{m} x_{ij} = b_j,\ i = 1,\ 2, \ldots, m; j = 1,\ 2, \ldots, n. \quad (1.4)$$

$$x_{ij} \geq 0 \text{ and an integer for all } i \text{ and } j. \quad (1.5)$$

Note C_{ij} denotes the per unit cost of transportation of a unit from the supply point i to the demand point j, a_i, and b_j denote supply at the warehouse i and demand at the destination j where $i = 1, 2, \ldots, m$ and $j = 1, 2, \ldots, n$.

Note that when all $a_i = 1$, $b_j = 1$, and $m = n$, the transportation model becomes an assignment model and similarly, if the row i in the transportation model is duplicated for a_i number of times and the column j is duplicated for b_j number of times with the supply and demand equal to 1, a balanced transportation model will become an assignment model of size $(M \times M)$. This is further explained in Section 1.4.

1.4 Consideration of the Transportation Model as an Assignment Model

As was pointed out above, if each row i in the transportation model is repeated for a_i number of times and similarly each column j is repeated for b_j number of times, each element C_{ij} in the transportation cost matrix can be seen as a submatrix of dimension a_i and b_j, as shown below:

$$\begin{bmatrix} C_{ij,11} & \cdots & C_{ij,1bj} \\ \cdots & \cdots & \\ C_{ij,ai\,1} & \cdots & C_{ij,aibj} \end{bmatrix}. \quad (1.6)$$

When each C_{ij} for all i and j is expanded in the form of a matrix as given by eqn (1.6), the new problem will be a $(M \times M)$ size assignment problem with demand and supply 1 at each supply and demand point, Here $M = \sum_{i=1}^{m} a_i = \sum_{j=1}^{n} b_j$. Since the new $(M \times M)$ size problem will be an assignment model, it is natural that the Hungarian method of assignment will be applicable and unlike the conventional transportation method, the problem can be solved as an assignment problem by using the Hungarian method of assignment or by the reconstruction approach. First, we make some observations in Section 1.4.1 and develop the Hungarian method for the transportation problem in Section 1.5.

1.4.1 Some observations

1.4.1.1 Observation 1

The Hungarian approach reconstructs equivalent cost matrices and attempts to find the assignment at zero total cost which is a natural minimum in the reconstructed cost structure of the network, hence is also optimal for the given cost data, as the relative merit of a technician in relation to a job was maintained, while cost matrix was reconstructed. This process of subtraction eliminates the need to prove the optimality of the solution.

1.4.1.2 Observation 2

The Hungarian method of assignment, which deals with the modified cost elements and after detecting an independent zero element, makes an allocation and draws a horizontal or a vertical line, to indicate no further assignments are possible in that row or column. However, in the case of transportation problem, if the requirement of such independent zero elements is represented by a number p, then the number p must satisfy the condition (1.7).

$$\max(m, n) \leq p \leq (m + n - 1)\,. \tag{1.7}$$

Note that $(m + n - 1)$ is the maximum number of allocations for a balanced transportation model. However, when for any row and column, if $a_i = b_j$; one single allocation will delete both, i.e., the ith row and the jth column by an allocation at the cell (ij). This will lead to a situation that a feasible solution can be reached in less than $(m+n-1)$ number of allocations, which is called a degenerate solution.

1.4.1.3 Observation 3

In a non-degenerate balanced transportation problem, the number of allocations will be given by $(m + n - 1)$. It means the number of horizontal lines must not exceed $(m - 1)$ and similarly the number of vertical lines also must not exceed $(n - 1)$ for a potential further modification of the relative cost elements C_{ij}. This observation is obvious, as when allocations result in 'm' number of horizontal lines, all rows have been crossed out, no element will be left as a positive uncrossed element for creating more zero entries. Similar will be the situation when all vertical lines are drawn for making allocations.

These m number of horizontal lines or n number of vertical lines will not allow further modification of the cost elements C_{ij}. After $(n - 1)$ vertical lines and $(m - 1)$ horizontal lines, only one element will be left as uncrossed element.

1.4.1.4 Observation 4

In a non-degenerate balanced transportation problem, the maximum number of independent zero elements that can be created by the Hungarian method of assignment will be at most $(m + n - 1)$, which is exactly the requirement for determination of the optimal solution of the transportation model. Hence the Hungarian method of assignment has a possible extension to solve a balanced transportation model. However, while allocations are made, the number of crossed out horizontal line will not exceed the number $(m - 1)$ and similarly the number of vertical lines should not exceed the number $(n - 1)$. The last allocation or the $(m + n - 1)$th allocation will lead to a feasible solution and at this stage either all rows or all columns will be crossed out. Thus, further creation of zero elements by the Hungarian method of assignment will not be possible.

The steps of the Hungarian method for the transportation model are discussed in Section 1.5.

1.5 Modified Hungarian Method of Assignment for Solving a Transportation Problem

The steps of the Hungarian Method of assignment for solving a balanced transportation model are as follows:

Step 1: Consider a balanced transportation model, as all transportation models can be easily converted to a balanced model by the introduction of an appropriate dummy supply or a demand point, there is no loss of generality.

Step 2: Subtract the minimum element in each row from all other elements in that row. This process will generate at least one zero element in each row. Repeat the same process for each column, thus once again, each column will have at least one zero element in each column in the given balanced $(m \times n)$ transportation cost matrix.

Step 3: Inspect row i, where $i = 1, 2, \ldots, m$, if the row i has a unique zero element, i.e., the modified $\overline{C_{ij}} = 0$ in column j, where $j = 1, 2, \ldots, n$; find the $\min(a_i,\ b_j)$.

(1) If $a_i = b_j$, draw a vertical line through the column j, as was done in the case of an assignment model. The modified value of a_i will be zero and similarly the modified value of b_j will also be zero. In other words, the $(m \times n)$ model will be reduced to a $((m-1) \times (n-1))$ model, with the row i missing and column j missing.

(2) If $a_i < b_j$, draw a horizontal line through the row i, if the number of existing horizontal lines is $<(m-1)$. Change the value a_i to zero, and the b_j value is changed to $(b_j - a_i)$. The remaining problem is reduced to $((m-1) \times n)$ dimension.

(3) If $a_i > b_j$, draw a vertical line in the column j, if the number of vertical lines is $<(n-1)$. Change the value of b_j to zero, and the value of a_i will be changed to $(a_i - b_j)$. The remaining problem will be reduced to $(m \times (n-1))$ dimension.

Repeat the same process for each row and after that for each column, and when all zero elements have been covered, go to Step 4.

If all zero elements have not been covered by a horizontal or a vertical line, find for each zero the possible maximum allocation and draw the line as per the rule described (1) to (3). Repeat this process until all zero elements have been covered. Go to Step 4.

Step 4: If the solution is feasible, go to Step 6, and if the solution is not feasible, and all zero elements in the modified matrix have been covered, do the following:If the total number of lines are less than $(m+n-2)$, find the minimum positive modified element, which has not been crossed out by a vertical or the horizontal line. Using the value of this minimum element, update the cost matrix as follows:

(i) Subtract this minimum value from all elements which have not been crossed out.

(ii) Do not alter values which are on a single horizontal or on a vertical line.

(iii) Add the minimum value to all elements which are on the intersection of two lines.

If the number of lines is equal to $(m+n-2)$, only one element will be left as an uncrossed positive number. If this number is large compared to other elements in that row or column, we get the feasible solution and go to Step 7.

Step 5: Go to Step 2, i.e., make sure once again that each row and column has a zero.

Step 6: Each allocation results in a horizontal or a vertical line, and if total number of lines is equal to $(m+n-2)$, then this last allocation must lead to a feasible solution, and that will be an optimal solution, if this relative cost element is minimum in that column or row, as may be the case. The process will terminate.

Step 7: If the last uncrossed element is not minimum in that row or column, obtain a feasible solution, and subject this feasible solution to the test of optimality by the standard (u, v) approach (i.e., dual feasible solution), and obtain the optimal solution by the conventional approach.

This modified Hungarian approach for the transportation model will be able to handle degenerate transportation models, without any difficulty.

1.6 Interesting Properties Associated with a Transportation Model

1.6.1 The transportation model property 1

Unlike a linear program, a special characteristic of a balanced transportation model is that any feasible solution can be tested for optimality without any past information.

Proof: The LP approach is sequential, for example, the extreme point search at any point of the LP convex space is dependent on the previous extreme point search. In other words, if an LP is given and its corresponding optimal solution is given, but the optimal simplex tableau is missing, we must resolve the problem to establish optimality of that solution. In the case of a transportation model, the optimality proof is based on values of the dual variables corresponding to the given feasible solution of that transportation model, which are directly linked to the values of the basic variables, Therefore, given a basic feasible solution, it can be tested for optimality at any time. This property is used in Step 7 of the algorithm in the previous section. Therefore, when the optimality of a solution is under any doubt, that solution can be subject to the test of the optimality.

1.6.2 The transportation model property 2

The Hungarian approach discussed in this chapter deals with a non-degenerate feasible solution comprised of $(m + n - 1)$ allocations at zero modified cost elements and the last allocation is at a positive non-zero modified cost element. This feasible solution obtained from the Hungarian approach to a transportation model can easily tested for optimality of the solution. However, the Hungarian approach is sequential like the LP, i.e., the process will have to start from the very beginning, if previous information is missing.

1.7 Numerical Illustrations

We present three illustrations, dealing with alternative, unique and degenerate solutions.

1.7.1 Example 1.1

Consider an example of a transportation problem taken from Eppen, Gould, and Schmidt (1993), page 283. The cost per unit, demand, and supply is given in Table 1.1.

Table 1.1 Given balanced transportation problem.

	1	**2**	**3**	**4**	**Supply**
1	12	13	4	6	500
2	6	4	10	11	700
3	10	9	12	4	800
Demand	400	900	200	500	2000

Note the minimum element in each row is 4. For creating a zero in each row, we subtract 4 from each element in rows 1, 2, and 3, and obtain Table 1.2 as given below.

Table 1.2 Reconstructed values of the unit transportation cost.

	1	**2**	**3**	**4**	**Supply**
1	8	9	0	2	500
2	2	0	6	7	700
3	6	5	8	0	800
Demand	400	900	200	500	2000

Since column 1 does not have a zero element, we again subtract 2 from the elements in column 1, and the costs per unit are further reconstructed as given in Table 1.3. Allocated values are shown in parenthesis and horizontal lines through a row are indicated by $\rightarrow$ and a vertical line through a column is indicated $\uparrow$ in the supply row and demand column, respectively. A number with an arrow indicates the order of the line. For example, in Table 1.3, column 3 has a vertical arrow with a number 1, it means the first allocation was made in this column in the cell (1,3).

Table 1.3 Reconstructed values of unit transportation cost.

	1	**2**	**3**	**4**	**Supply**
1	6	9	0 (200)	2	~~500~~, 300
2	0 (400)	0 (300)	6	7	~~700~~, 300, 0→ 4
3	4	5	8	0 (500)	~~800~~, 300
Demand	~~400~~,0 ↑ 3	~~900~~, 600	~~200~~,0 ↑1	~~500~~, 0 ↑ 2	~~2000~~, 600

Note that columns 1, 3, and 4 have been crossed out by vertical lines and row 2 by a horizontal line and the solution is not yet feasible, as 600 more units must be allocated. We need to create a greater number of zero elements. Among the two not crossed-out elements in column 2, the minimum is 5, situated in the cell (3,2). Reconstructed values are given in Table 1.4.

Table 1.4 Reconstructed values of the unit transportation cost from Table 1.3.

	1	**2**	**3**	**4**	**Supply**
1	6	4	0	2	500
2	5	0	11	12	700
3	4	0	8	0	800
Demand	400	900	200	500	2000

Note that column 1 has no zero elements, hence subtracting the minimum value 4 from all elements in this column, a zero element is created as shown in Table 1.5.

Table 1.5 Reconstructed values of the unit transportation cost.

	1	**2**	**3**	**4**	**Supply**
1	2	4	0 (200)	2	~~500~~, 300
2	1	0 (700)	11	12	~~700~~, 0→ 2
3	0 (400)	0(200)	8	0 (200)	~~800~~, ~~400~~, ~~200~~ 0→ 5
Demand	~~400~~,0 ↑ 3	~~900~~, ~~200~~ 0↑ 4	~~200~~,0 ↑1	~~500~~, 300	~~2000~~, 300

The solution is still infeasible, hence more zeros must be created. Columns 1, 2, and 3 have been covered by vertical lines, rows 2 and 3 by the horizontal lines. Hence the only element not crossed out is in the cell (1,4). Its value is 2, which is minimum element in row 1 and column 4 that has no allocation. The new updated values are shown in Table 1.6.

Table 1.6 Reconstructed values of the unit transportation cost.

	1	**2**	**3**	**4**	**Supply**
1	2	4	0	0	500
2	3	2	13	12	700
3	2	2	10	0	800
Demand	400	900	200	500	2000

Now columns 1 and 2 have no zero elements, subtracting the minimum, and making allocations, we obtain Table 1.7.

Table 1.7 Reconstructed values of the unit transportation cost from Table 1.6.

	1	**2**	**3**	**4**	**Supply**
1	0	2	0 (200)	0 (300)	~~500~~, ~~300~~, 0 0→ 6
2	1	0 (700)	13	12	~~700~~, 0 0→ 1
3	0 (400)	0 (200)	10	0 (200)	~~800~~, ~~600~~, ~~200~~, 0 0→ 7
Demand	~~400~~ 0↑4	~~900~~, ~~200~~ 0↑ 2	~~200~~ 0↑ 3	~~500~~,~~300~~, 0	~~2000~~, 0

In Table 1.7, we have two zero elements in rows 1 and 3 and two zeros in columns 1 and 4, hence more than one optimal solution may exist. One possible solution is displayed in Table 1.7, which is optimal and total cost is $12,000. The second alternate optimal solution is shown in Table 1.8.

Table 1.8 The alternate optimal solution.

	1	**2**	**3**	**4**	**Supply**
1	12 (300)	13	4(200)	6	500
2	6	4(700)	10	11	700
3	10 (100)	9(200)	12	4(500)	800
Demand	400	900	200	500	2000

1.7.2 Example 1.2

Consider one more transportation problem taken from Munapo and Kumar (2022), page 132 as given below in Table 1.9.

Table 1.9 The transportation model.

	1	**2**	**3**	**4**	**5**	**6**	**Supply**
1	8	9	20	4	4	6	**20**
2	2	1	1	1	1	1	**30**
3	2	8	19	7	3	14	**40**
4	2	6	6	4	11	8	**50**
5	2	15	5	12	8	7	**60**
Demand	30	**40**	**50**	**20**	**20**	**40**	**200**

After subtracting 4 from the elements in row 1, 1 from the elements in row 2, and 2 from the elements in rows 3, 4, and 5, the reconstructed cost matrix will be as given in Table 1.10.

Table 1.10 With at least one zero in each row.

	1	**2**	**3**	**4**	**5**	**6**	**Supply**
1	4	5	16	0	0	2	**20**
2	1	0	0	0	0	0	**30**
3	0	6	17	5	1	12	**40**
4	0	4	4	2	9	6	**50**
5	0	13	3	10	6	5	**60**
Demand	30	**40**	**50**	**20**	**20**	**40**	**200**

Note that each column in Table 1.10 has a zero element, hence no further reconstruction of costs will be necessary. Following the allocation procedure, we get allocations as shown in Table 1.11.

Table 1.11 Making allocation in cells with modified cost 0.

	1	**2**	**3**	**4**	**5**	**6**	**Supply**
1	4	5	16	0 (20)	0	2	**~~20~~**, 0 → 3
2	1	0 (30)	0	0	0	0	**~~30~~, 0, → 2**
3	0 (30)	6	17	5	1	12	**~~40~~, 10**
4	0	4	4	2	9	6	**50**
5	0	13	3	10	6	5	**60**
Demand	~~30~~ 0↑1	**~~40~~, 10**	**50**	**~~20~~,0**	**20**	**40**	**~~200~~, 120**

The solution in Table 1.11 is not feasible; hence, we must reconstruct the cost elements in Table 1.11 following Step 4, which is shown in Table 1.12. The minimum non-zero element, which has not been crossed out is 1 in cell (3,5). Note the horizontal lines are in rows 1 and 2 and the vertical line is in column 1.

Table 1.12 Reconstructed cost elements are shown below.

	1	**2**	**3**	**4**	**5**	**6**	**Supply**
1	5	5	16	0 (20)	0	2	~~20~~, 0 → 3
2	2	0 (30)	0	0	0	0	**~~30~~, 0 →2**
3	0	5	16	4	0 (20)	11	**~~40~~, 20**
4	0(30)	3	3	1	8	5	**~~50~~, 20**
5	0	12	2	9	5	4	**60**
Demand	**~~30~~** 0↑ 1	**~~40~~,10**	**50**	**~~20~~, 0**	**~~20~~,** 0↑ 4	**40**	**~~200~~ 100**

The solution in Table 1.12 is still infeasible. The minimum uncrossed element is 1 in the cell (4,4). Reconstructed cost values are updated and given in Table 1.13.

Table 1.13 Updated elements from Table 1.12 are shown below.

	1	**2**	**3**	**4**	**5**	**6**	**Supply**
1	6	5	16	0 (20)	1	2	~~20~~, 0
2	3	0 (30)	0	0	1	0	**~~30~~, 0→ 3**
3	0	4	15	3	0 (20)	10	**~~40~~, 20**
4	0(30)	2	2	0	8	4	**~~50~~, 20**
5	0	11	1	8	5	3	**60**
Demand	**~~30~~** 0↑ 2	**~~40~~, 10**	**50**	**~~20~~,** 0↑ 1	**~~20~~,** 0↑ 4	**40**	**~~200~~,100**

Solution is not feasible. Cost values need further reconstruction. Note the minimum uncrossed element is 1 in cell (5,3), which will be used to update the elements as shown in Table 1.14.

Table 1.14 Reconstructed cost values from Table 1.13 are shown below.

	1	**2**	**3**	**4**	**5**	**6**	**Supply**
1	6	4	15	0 (20)	1	1	~~20~~, 0
2	4	0 (30)	0	1	2	0	**~~30~~, 0 → 3**
3	0	3	14	3	0 (20)	9	**~~40~~, 20**
4	0(30)	1	1	0	8	3	**~~50~~, 20**
5	0	10	0 (50)	8	5	2	**~~60~~, 10**
Demand	**~~30~~** 0 ↑ 2	**~~40~~, 10**	**~~50~~** 0↑ 5	**~~20~~,** 0 ↑ 1	**~~20~~,** 0↑ 4	**40**	**~~200~~,50**

Once again, the solution is not yet feasible. Note the minimum of those elements which have not been crossed out is 1 in cell (4,2). The reconstructed cost information is given in Table 1.15.

Table 1.15 Updated elements from Table 1.14 are shown below.

	1	**2**	**3**	**4**	**5**	**6**	**Supply**
1	6	3	15	0 (10)	1	0 (10)	~~20~~, ~~10~~ **0→ 9**
2	5	0	1	2	3	0 (30)	**30,** **0 → 6**
3	0 (20)	2	14	3	0 (20)	8	**~~40, 20,~~** **0 →3**
4	0	0 (40)	1	0 (10)	8	2	**~~50, 10.~~** 0 → **7**
5	0 (10)	9	0 (50)	8	5	1	**~~60~~, 10,** **0**
Demand	~~30,~~ ~~10~~ 0 ↑ 4*	**~~40~~,** **0** ↑ 5*	**~~50~~** 0 ↑ 1*	**~~20~~,** 0 ↑ 8*	**~~20~~,** 0 ↑ 2*	**~~40~~** **20**	**~~200~~, 0**

The solution in Table 1.15 is feasible and hence optimal. Note that the optimal solution stated by Munapo and Kumar (2022) is 780, which is the same as obtained by the modified Hungarian method for the transportation model. It is a degenerate solution, where the test of optimality by the conventional approach will be more challenging.

1.7.3 A degenerate transportation model

We now consider a degenerate transportation model taken from Sharma (2018) as given in Table 1.16.

Example 1.3

Table 1.16 Degenerate transportation model.

	D_1	D_1	D_1	**Supply**
S_1	8	5	6	120
S_1	15	10	12	80
S_1	3	9	10	80
Demand	150	80	50	280

After creating a zero element in each row and column, we get from Table 1.16, Table 1.17, as given below with allocations.

Since the solution in Table 1.17 is infeasible, cost elements must be reconstructed again and more zero elements must be created. Vertical lines are in columns 2 and 3, and a horizontal line is in row 3. The minimum of the

Table 1.17 With modifies per unit transportation cost.

	D_1	D_1	D_1	**Supply**
S_1	3	0	0 (50)	~~120~~,70
S_1	5	0(80)	1	~~80~~, 0
S_1	0 (80)	6	6	~~80~~,0 $\rightarrow$ 2
Demand	~~150~~,70	~~80~~,0 $\uparrow$ 1	~~50~~,0, $\uparrow$ 3	~~280~~, 70

element not crossed out is 3 in the cell (1,1). The updated values are given in Table 1.18.

Table 1.18 Reconstructed cost values from Table 1.17.

	D_1	D_1	D_1	**Supply**
S_1	0 (70)	0	0 (50)	~~120~~,~~70~~, 0, $\rightarrow$ 4
S_1	2	0(80)	1	~~80~~, 0
S_1	0 (80)	9	9	~~80~~,0 $\rightarrow$ 2
Demand	~~150~~,~~70~~, 0	~~80~~,0 $\uparrow$ 1	~~50~~,0, $\uparrow$ 3	~~280~~, ~~70~~, 0

The solution in Table 1.18 is feasible and hence optimal. Once again, it is the same solution as given by Sharma (2018). Normally for this problem, the test for optimality will require 5 allocations, but by the modified Hungarian method for the transportation problem, degeneracy does not cause any difficulty for establishing an optimality of the solution.

1.8 Accelerating the Search Process for the Hungarian Method of Assignment

1.8.1 Complexity and weakness of the Hungarian method of assignment

The Hungarian method has many versions of the original algorithm, and its complexity $O(n^4)$ was established by Munkers (1957). Later it was shown that the complexity of the Hungarian approach is of the order $O(n^3)$ as discussed by Edmonds and Karp (1972) and Tomizawa (1971). Recently Munapo (2020) pointed out that the Hungarian method of assignment can be slow and search process can be accelerated. In the Hungarian approach, when the number of independent zeros is insufficient, we look for the smallest uncovered elements, which may be only one, and thereby one more zero element is created, thus progress towards the optimal solution can be very slow. Creating a single zero at each iteration becomes a weakness of the Hungarian method. Creating more than one zero is like accelerating the search process. Munapo (2020) suggested that cost elements for a given assignment problem can be changed without changing the relative merits of costs associated with various combinations and possibilities.

Consider the cost matrix as given in Table 1.19.

Table 1.19 A general assignment cost matrix.

(c_{11})	(c_{12})	$\dots$	(c_{1n})
(c_{21})	(c_{22})	$\dots$	(c_{2n})
$\dots$			$\dots$
(c_{n1})	(c_{n2})	$\dots$	(c_{nn})

Cost data in Table 1.19 can be changed to cost data in Table 1.20, without changing the relative merits of various assignments, i.e., the optimal solution in Table 1.20 will maintain its optimality with respect to the cost data in Table 1.19. Therefore, for determination of the optimal solution, we use the modified data in Table 1.20.

Table 1.20 Modified cost data without changing relative merits.

$(c_{11} + \alpha_1)$	$(c_{12} + \alpha_2)$	$\dots$	$(c_{1n} + \alpha_n)$
$(c_{21} + \alpha_1)$	$(c_{22} + \alpha_2)$	$\dots$	$(c_{2n} + \alpha_n)$
$\dots$			$\dots$
$(c_{n1} + \alpha_1)$	$(c_{n2} + \alpha_2)$	$\dots$	$(c_{nn} + \alpha_n)$

This fact can be used to create a column or a row of any balanced transportation problem to a same number and thereby creating a column or a row of all zeros. This is illustrated by using the cost data from Example 1.4, for which see the cost data given in Table 1.21.

1.8.2 Illustration of acceleration step for the optimal search: Example 1.4

Table 1.21 Cost data for an illustration.

Job/worker	1	2	3	4	5	6
1	11	9	4	10	25	14
2	15	25	3	33	19	13
3	45	17	3	23	16	20
4	3	3	4	2	2	2
5	32	33	5	44	29	42
6	26	63	3	30	47	23

Here the maximum in column 1 is 45. If we add appropriate numbers in each row, this column can be converted to have the same cost value, i.e., 45. For example, consider the following operation that results in Table 1.22. Alter the cost data in row 1 is altered by adding the number $(r_1 + 34)$, row two by the number $(r_2 + 30)$, row 3 by the number $(r_3 + 0)$, row 4 by the number $(r_4 + 42)$, row 5 by the number $(r_5 + 13)$, and finally row 6 is altered by $(r_6 + 19)$, here r_i are the cost elements in row i, where $i = 1, 2, \ldots, 6$. This results in Table 1.22.

Table 1.22 Altered cost data without changing relative merits.

Job/worker	1	2	3	4	5	6
1	45	43	38	44	59	48
2	45	55	33	63	49	43
3	45	17	3	23	16	20
4	45	45	46	44	44	44
5	45	46	18	57	42	55
6	45	82	22	49	66	42

Ensuring that every column has a zero, we subtract 45 from the elements of column 1, 17 from the elements of column 2, 3 from the elements of column 3, 23 from the elements of column 4, 16 from the elements of column 5 and 20 from the elements of column 6. This gives rise to cost data given in Table 1.23. We find the non-zero minimum for each column, which is given in Table 1.23. We make these minimum equal to the maximum of these minimums, i.e., 26.

Table 1.23 At least one zero in each column.

Job/worker	1	2	3	4	5	6	
1	0	26	35	21	43	28	
2	0	38	30	40	33	23	
3	0	0	0	0	0	0	$\rightarrow 2$
4	0	28	43	21	28	24	
5	0	29	15	34	26	35	
6	0	65	19	26	50	22	
Non-zero min	$\uparrow 1$	26	15	21	26	22	
Add to		0	11	5	0	4	

These zero entries can be covered by two lines. For the transformed cost matrix in Table 1.24, we apply the Hungarian step of creating more zeros. The minimum is 26. We obtain the cost data in Table 1.24.

Table 1.24 Modified cost data, not changing any relative merit.

Job/worker	**1**	**2**	**3**	**4**	**5**	**6**	
1	0	0 A	20	0	17	6	→ 4
2	0 A	12	15	19	7	1	
3	26	0	11	5	0A	4	→5
4	0	2	28	0A	2	2	→2
5	0	3	0A	13	0	13	→ 6
6	0	39	4	5	24	0A	
Add to	↑ 1	0	1	↑ 2	0	↑3	

Therefore, in two iterations, we have the following optimal solution. It is (2,1), (4,4), (6,6), (1,2), (3,5), and (5,3). The optimal cost can be easily obtained from the cost data given in Table 1.22.

1.8.3 Steps for the accelerated Hungarian method of assignment

Assuming that we are dealing with a balanced assignment problem. Algorithmic steps are as follows:

Step 1: Create a 'all-zero' row or a 'all-zero' column and ensure that all other rows and columns have at least one zero entry. This can be achieved by adding an appropriate number to each row or column.

Step 2: Cover all zeros with a minimum number of lines. If the number of lines is equal to 'n', then an optimal solution can be identified, and search process comes to an end. Else, go to Step 3.

Step 3: If the number of lines is less than n, and all zero elements have been covered. Select the smallest uncovered element and bring all other smaller elements to the same level to create more than one zero in one iteration of the usual Hungarian steps, i.e.,

1. Element on one line remains unchanged.
2. Minimum element is added to an element on the double lines.
3. Minimum element is subtracted from all uncrossed elements.

Return to Step 2.

1.8.4 Hungarian method of assignment for the transportation model

It has already been identified that balanced transportation problem can be seen as an assignment problem, and three illustrations were given in Section 1.7. The Hungarian method for a given ($m \times n$) transportation model can be slightly modified to take advantage of the acceleration idea by following the steps given below.

Step 1: Identify the value of *m* and *n* for the given problem.

1.1 If $m = n$, select the row or the column with $\max\left[\{\max\left(a_i\right)\},\ \{\max\left(b_j\right)\}\right]$.
1.2 If $m < n$, select the row with $\max\left(a_i\right)$.
1.3 If $m > n$, select the column with $\max(b_j)$.

Step 2: In the selected row or the column, by adding appropriate numbers create 'all-zero' entries and ensure that all other rows and columns have at least one zero entry.

Step 3: Cover all zeros with a minimum number of lines and allocate feasible quantity at a zero entry one for each line. If the solution is feasible, then an optimal solution can be identified, and search process comes to an end. Else, go to Step 4.

Step 4: If the solution is not feasible and all possible allocations have been made at zero elements, which have been covered. It means more zero element cells will be required. Select the smallest uncovered element and bring all smallest elements in the uncovered rows or column to the same level to create more than one zero following the usual Hungarian steps, i.e.,

1. Element on one line remains unchanged.
2. Minimum element is added to an element on the double lines.
3. Minimum element is subtracted from all uncrossed elements.

Return to Step 2.

We reconsider the three illustrations discussed in Section 1.7 again and illustrate the acceleration idea on transportation models providing only partial solution. We will, once again, obtain the same solutions obtained earlier.

1.8.5 Example 1.5

Let us reconsider:

Table 1.25 Cost data from Example 1.1.

	1	**2**	**3**	**4**	**Supply**
1	12	13	4	6	500
2	6	4	10	11	700
3	10	9	12	4	800
Demand	400	900	200	500	2000

Note that $n = 4$, and $m = 3$; hence, the $n > m$. The maximum supply corresponds to row 3, which is selected for creating all zero entries. The maximum cost value in row 3 is 12. We make all elements in row 3 equal to 12, i.e., by adding 2 to all elements in column 1, 3 to all elements in column 2, and 8 to all elements in column 4, resulting in Table 1.26.

Table 1.26 Modified cost values from Table 1.25.

	1	**2**	**3**	**4**	**Supply**
1	14	16	4	14	500
2	8	7	10	19	700
3	12	12	12	12	800
Demand	400	900	200	500	2000

Creating at least one zero in each row, we have Table 1.27.

Table 1.27 Modified cost values with at least one zero in each row.

	1	**2**	**3**	**4**	**Supply**
1	10	12	0 (200)	10	500
2→ 2	1	0(700)	3	12	700
3→ 5	0(400)	0(200)	0	0(200)	800
Demand	400 ↑ 3	900 ↑ 4	200 ↑ 1	500	2000

Note the minimum uncrossed element is 10 and it is the only uncrossed element. No further preparation is necessary, and we apply the usual rule of Hungarian method to update the cost elements and get Table 1.28.

Table 1.28 Modified cost values from Table 1.27.

	1	**2**	**3**	**4**	**Supply**
1	10	12	0	0	500
2	11	10	13	12	700
3	10	10	10	0	800
Demand	400	900	200	500	2000

Again, making sure of a zero element in each column, we have Table 1.29.

Table 1.29 Modified cost values from Table 1.28.

	1	**2**	**3**	**4**	**Supply**
1	0	2	0(200)	0(300)	500
2→ 1	1	0(700)	13	12	700
3→ 5	0(400)	0(200)	10	0(200)	800
Demand	400 ↑ 3	900 ↑ 4	200 ↑ 2	500 ↑ 6	2000

Clearly, the above solution is optimal, and one more solution exists for this problem.

1.8.6 Example 1.6

Here we reconsider Example 1.2 and illustrate the accelerating step for this transportation model. The cost data is given in Table 1.30.

Table 1.30 The transportation model.

	1	**2**	**3**	**4**	**5**	**6**	**Supply**
1	8	9	20	4	4	6	**20**
2	2	1	1	1	1	1	**30**
3	2	8	19	7	3	14	**40**
4	2	6	6	4	11	8	**50**
5	2	15	5	12	8	7	**60**
Demand	30	**40**	**50**	**20**	**20**	**40**	**200**

In this case, since $n > m$, we create a zero-entry row which will be row 5. For that we add 13 to all elements in column 1, no change to column 2, 10 is added to all elements in column 3, 3 is added to all elements in column 4, 7 is added to all elements in column 5 and 8 is added to all elements in 6. This results in Table 1.31.

Table 1.31 Modified cost data for the transportation model.

	1	**2**	**3**	**4**	**5**	**6**	**Supply**
1	21	9	30	7	11	14	**20**
2	15	1	11	4	8	9	**30**
3	15	8	29	10	10	22	**40**
4	15	6	16	7	18	16	**50**
5	15	15	15	15	15	15	**60**
Demand	30	**40**	**50**	**20**	**20**	**40**	**200**

We can either make the minimum in each row equal or directly creating a zero entry in each row, we obtain Table 1.32.

Table 1.32 The transportation model.

	1	**2**	**3**	**4**	**5**	**6**	**Supply**
1	14	2	23	0(20)	4	7	**20**
2	14	0(30)	10	3	7	8	**30**→ 2
3	7	0(10)	21	2	2	14	**40**
4	9	0	10	1	12	10	**50**
5	0(30)	0	0(30)	0	0	0	**60**→ 5
Demand	30 ↑ 4	**40** ↑ 3	**50**	**20** ↑ 1	**20**	**40**	**200/80**

Note that in the accelerated approach, 120 units have been allocated. The optimal solution will take a few more iterations as was shown earlier in Example 1.2.

Before we conclude this chapter, we apply one more time acceleration idea to the cost matrix for Example 1.3, which is reproduced below as Table 1.33.

1.8.7 Example 1.7

Table 1.33 Degenerate transportation model.

	D_1	D_1	D_1	**Supply**
S_1	8	5	6	120
S_1	15	10	12	80
S_1	3	9	10	80
Demand	150	80	50	280

Here *m* and *n* are equal, hence we select column 1 for all zero entries by bringing all cost elements in this column to 15. This can be seen as given in Table 1.34.

Table 1.34 Creation of all zero column.

	D_1	D_1	D_1	**Supply**
S_1	15	12	13	120
S_1	15	10	12	80
S_1	15	21	22	80
Demand	150	80	50	280

Subtract minimum in each column, in Table 1.35, below.

Table 1.35 All zero-entry column.

	D_1	D_1	D_1	**Supply**
S_1	0	2	1	120
S_1	0	0	0	80
S_1	0	11	12	80
Demand	150	80	50	280

Finally bringing minimum at the same level, i.e., by adding 1 to elements in column 3, we have Table 1.36.

Table 1.36 Acceleration idea applied to column 3.

	D_1	D_1	D_1	**Supply**
S_1	0	2	2	120
S_1	0	0	1	80
S_1	0	11	13	80
Demand	150	80	50	280

1.9 Concluding Remarks

In this chapter the transportation network has been reconstructed with respect to link weights and a modified Hungarian method has been developed for solving the transportation model. The complexity of the assignment problem has been discussed by Dantzig (2002).This idea was discussed by Kumar et al. (2023). In this approach, the optimality of any feasible solution is independent of the order of degeneracy. Several problems have been solved to illustrate various situations. Later, a recent idea of acceleration developed by Munapo (2020) has been also discussed and modified for an application to the transportation model.

Testing the optimality of a feasible solution is easy for the transportation model by the dual optimality test. It can easily be applied to any feasible transportation solution, if required. Optimality test is required only when the solution is comprised of $(m + n - 1)$ allocations and the last allocation element does not correspond to a minimum in that row or the column.

It may also be noted that since a transportation model has been considered as an assignment model, we can apply ideas to find the kth best of the assignment to the transportation model, where $k \geq 2$. Please see Murthy (1968) and Kumar et al. (2020, 2023).

References

H. Aboli, N. M. Parikshit, Trends and Challenges in Measuring Performance of Reviewer Paper Assignment, Procedia Computer Science, Volume 171, 709-718, 2020.

G. B. Dantzig, Linear programming, Operations Research, Vol 50, No. 1, pp 42-47, 2002.

J. Edmonds, and R. M. Karp, Theoretical improvements in Algorithmic efficiency for network flow problems, Journal of the ACM, 19(2) 248-264, 1972, doi: https://doi.org/10.1145/321694.3211699

G. D. Eppen, F. J. Gould, C. P. Schmidt, Introduction to Management Science, Fourth Edition, Prentice Hall, New Jersey, 1993.

F. S. Hillier, G. J. Lieberman, Introduction to Operations Research, 2015, ISBN 13:79812 59545962

H. W. Kuhn, The Hungarian method of assignment problem, Naval Research Logistics Quarterly, 2, pp 83-97, 1955.

H. W. Kuhn (Republished), The Hungarian method of assignment problem, Naval Research Logistics Quarterly, 52(1), 7-21, 2005.

S. Kumar, O. Ncube, E. Munapo, Tosoro and Hungarian approaches: A Hybrid algorithm for an assignment problem, Asia-Pacific Journal of Operational Research, Vol 20, pp 41-49, 2003.

S. Kumar, Ali Al-Hasani, M. Al-Rabeeah, and A. Ebehard, A random search method for finding "k>=2" number of ranked optimal solution to an assignment problem, Journal of Physics, Conference series 1490(2020)012063, pp 1-13, 2020. https://doi:10.1088/1742-6596/1490/1/012063.

S. Kumar, T. Tawanda, E. Munapo, and P. Nyamugure, A note on solving the transportation problem by the Hungarian method of Assignment: A total unification of transportation and Assignment models, In Intelligent Computing and Optimization, Vol 5, Springer, Eds. P. Vasant et al., Lecture notes in Network Systems, 855, 2023, pp 301-313, htti://doi.org/10.1007/978-3-031-50158-6

E. Munapo, P. Nyamugure, M. Lesaoana, and S. Kumar, A note on unified approach to solving transportation and assignment models in Operations Research, International Journal of Mathematical Modelling, Simulation and Applications, Vol 5, No. 2, pp 140-149, 2012.

E. Munapo, S. Kumar, Linear Integer Programming: Theory, Applications, Recent Developments, De Gruyter Series on the Applications of Mathematics in Engineering and Information Sciences, 2022, ISBN 978-3-11-070292-7

E. Munapo, Development of an accelerating Hungarian Method for assignment problems, Eastern-Europe Enterprise Technologies, 2020, DOI https://doi10.15587/1729/-4061.2020.209172

J. Munkers, Algorithms for the assignment and transportation problems, Journal of the Society for Industrial and Applied Mathematics, 5 (1), 32-38, 1957, doi: https://doi.org/10.1137/0105003

K. G. Murthy, An algorithm for ranking all the assignments in order of increasing costs, Operations Research, 16.3, pp 682-687, 1968.

J. K. Sharma, Operations Research: Theory and Applications, Trinity Press, 2018.

H. A. Taha, Operations Research: An Introduction, Pearson Educators, 2017, 10th Edition.

N. Tomizawa, On some techniques useful for solution of transportation network problems, Networks 1, 173-194, 1971. Doi: https://doi.org/10.1002/net.3230010206

W. L. Winston, Operations Research Applications and Algorithms, Duxbury Press, 4th Edition, 2004.

R. Zhang, M. Wang, and X. Pan, New model of the storage location of assignment problem considering demand correlation pattern, Computer, and Industrial Engineering 129, 210-219, 2019.

2

Reconstruction Approach for the Shortest and Kth Shortest Routes in Networks

Abstract

In this chapter shortest route problem has been analysed by reconstruction of link weights in a directed as well as non-directed network. The actual process of link weight modification for a directed network and non-directed network differs due to additional information about directions in directed networks. Reconstruction is a concept that requires designing of the reconstruction to suit a given situation. In this chapter, reconstruction approaches have been applied to network routing problems. We also introduce a concept of virtual directions and use that concept to find the shortest path in a non-directed network. We explain the purpose, process, and the philosophy of link weight reconstruction to find the shortest route in a directed and a non-directed network. The shortest routes in a network have various industrial applications. As an application, the concept of virtual directions has been applied to a probability non-directed network. A new concept on developing an alternative for the optimal solution has been developed and applied for solving the Kth best route in a network and assignment has been discussed in this chapter.

Keywords: Shortest route, directed networks, non-directed networks, virtual directions, link-weight reconstruction, Kth best assignment, Kth best route.

Notation

1. *G(n,m)* represents a network, where the set of nodes is represented by '*n*' and the set of links joining various node is represented by the set '*m*'.

2. (i, j) represents a link, if it is directed, it starts at node i and ends at node j, ($i < j$), for non-directed network, it is valid in both directions $(i, j = 1, 2, \ldots, n)$.
3. w_{ij} = the weight associated with the link (i, j).
4. L_0 = set of labelled nodes.
5. L_1 = set of links connecting the labelled nodes to unlabelled nodes.
6. L_2 = set of links that will never participate in the shortest path.
7. $L_{3(1,k)}$ = set of links that form the shortest path between the nodes '1' and the node k, $k = 1, 2,\ldots,n$.
8. Kth shortest route, $K >= 2$.

2.1 Introduction

Several situations can be represented by a network diagram *G(n,m)*, where '*n*' represents the set of nodes and '*m*' denotes the set of the links; each link in the set '*m*' joins a pair of nodes in that network. These nodes represent some physical entities and links represent the relationship among those entities. The relationship between the two entities is measured in the form of a weight assigned to that link. In many different fields of science, engineering, technology, management, business, and commerce, we often come across situations represented by a network and these problems are analysed by using some network methodology. Many of these real-life situations require determination of a shortest route between two given nodes, known as the origin node and the destination node. These network routing problems often arise in many real-life situations and the objective is to find the shortest route between two given nodes and the associated minimum weight with that route. The total weight associated with a route is the sum of link weights forming that route. A large volume of literature exists in operations research, where many real-life situations have been conceptually represented by a network; analysed as a network and the results have been interpreted back regarding the original situation, see Hastings (1973), Saksena (2007). They have considered real-life industrial situations that give rise to a directed network. They have analysed these networks by using the dynamic programming technique by Bellman (1957), and Bellman and Dreyfus (1962). Dynamic programming is a useful technique for the directed network and this approach gets cumbersome when applied to non-directed networks, see Bellman (1958).

The links in the networks are assigned weights to represent the object in a particular physical situation. These network models, depending on the physical situation, can be a directed or non-directed network and objective can be to maximize or minimize the given objective function under some given conditions. In the shortest route problem, the purpose is to find a path joining two specific points, known as the origin node and the destination node, such that the total weight of the links forming the path between the required two given points is minimum or the maximum as may be suited in that situation.

Determination of the shortest route in the network *G(n,m)* is a classical problem that has been tackled by several researchers. This classical routing problem has been solved by different researchers and many solution procedures were developed, see Bellman (1958), Kalaba (1960), Pert, Randolf, Battlett (1960), Saksena and Kumar (1966) a, Pollak and Weibenson (1960), Dantzig (1960), and Dijkstra (1959). Pollak and Weibenson (1960) presented a review of shortest routing problems.

This chapter presents a new approach, where the associated link weights are reconstructed to get to the required optimum solution. The concept of link weight modification has been around, as was pointed out in Chapter 1 but here in this chapter reconstruction of link weights has been explored in context with the network routing and assignment problems.

The chapter has been organized in 8 sections. Each concept is explained by giving a numerical illustration. Sections 2.2 and 2.3 deal with determination of the shortest route in a directed network. Sections 2.4 and 2.5 discuss the case of non-directed network and the concept of 'Virtual directions' in a non-directed network. In Section 2.6, an application of virtual directions has been used to find the maximum reliability route in a probability network. Ordered optimum solutions have been discussed in Section 2.7. Finally, the chapter has been concluded in Section 2.8 with some remarks and future challenges.

2.2 Shortest Route in a Directed Network

A directed network *G(n,m)*, discussed in this chapter assumes the following:

1. Link (i, j) is a directed link, starts at node i and ends at node j, ($i < j$), in other words nodes are numbered in a topological order, $i, j = 1, 2, \ldots, n$.
2. All nodes, except the source and the destination node, have at least one in-coming link to it and at least one out-going link from it.
3. The source has all out-going links and the destination node has all in-coming links. Without any loss of generality, node '*1*' is assumed to

denote the origin node and the last node '*n*' represents the destination node.

In this chapter, we introduce a concept of a link weight reconstruction. For a directed network reconstruction can be carried out as follows:

1. In a directed network, find the minimum incoming weight to a node. Subtract the minimum weight from the weight of all incoming links and add the same minimum weight to the weight of all outgoing links from that node. If the required optimal path passes through this node, the total length between the source node and the destination node does not change. This process can be implemented for changing link weights and to find the optimal path. This rule can be iteratively implemented on all nodes, and if a path joining the origin and the destination nodes can be identified, which passes through the links with reconstructed zero weights, resulting in a total associated cost to be zero, hence it will become the required minimum path. Note all link weights remain non-negative.
2. Each iteration results in the reconstruction of all links associated with a node under study. The reconstruction process will result in at least one link with zero link weight and link weights of all outgoing links will be modified by adding the same weight. Starting the reconstruction process at the origin node or from node 1, the process can be carried out at node $2, 3, \ldots,$ etc., until we reach the destination node '*n*'. In an '*n*' node network, number of iterations will be *n*.
3. A label (m, w) is associated with node *K*, where $K = 1, 2, \ldots, n$. The node m represents sequential position of the node from where we arrived to node *K* in the network and w represents the minimum weight from the source node to node *K*.

2.2.1 Illustration of link weight reconstruction

Consider the network in Figure 2.1, where *K* is a typical node in the network with three incoming links and three out-going links. Let w_1, w_2, w_3 be weights associated with the incoming links and let w_4, w_5, and w_6 be the weights associated with the out-going links. When minimum of the in-coming weights is determined at node *K*, that minimum will be the minimum of the weights (w_1, w_2, w_3). Let us assume that the minimum weight is w_2.

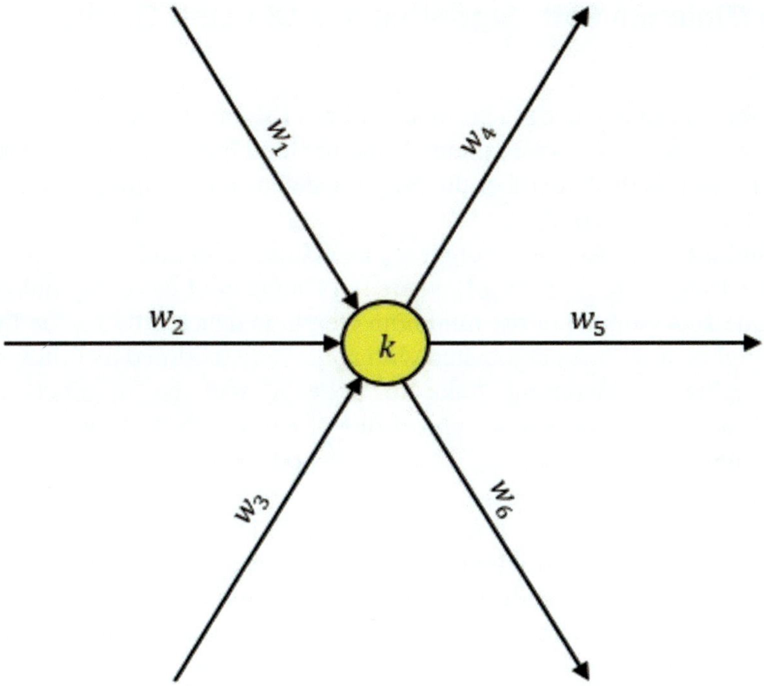

Figure 2.1 A typical illustration of weight reconstruction associated with the node.

Reconstructed weights will be non-negative and at least one of them will be zero. These reconstructed weights for the in-coming links will become $(w_1 - w_2)$, $(w_2 - w_2) = 0$, and $(w_3 - w_2)$. The weight associated with the out-going links will be $(w_4 + w_2)$, $(w_5 + w_2)$, and $(w_6 + w_2)$. Note all reconstructed weights are non-negative, i.e., $>= 0$. The above reconstructed weights assume that w_2 is the minimum incoming weight. One typical possibility is shown in Figure 2.2, total link weight using the links w_1 and w_4 remains unchanged, i.e., $w_1 + w_4$. Similar will be the case for all other eight possibilities.

Figure 2.2 A typical path with altered weights.

2.3 The Minimum Path Algorithm in a Directed Graph

Algorithmic steps are:

Step 1: Label the source node. This label will be (1,0) indicating that we start from node 1 and arrive at node 1 and the total link weight from node 1 to itself is 0. Remember the origin node has no incoming links. Set $k \to 2$, go to step 2.

Step 2: Find the min incoming weight w_k associated with node k, i.e. $w_k = \min\{w_{l1k}, w_{l2k}, \ldots, w_{lkk}\}$, where lk is number of incoming links to node k. Assume that the minimum weight is denoted by w, the link weights of all links associated with node k are modified as follows: Weights of incoming links to node k will be modified as: $(w_{lik} - w)$, and the weight of one of these links will be $= 0$. Weights for the outgoing links from node k will be: $(w_{kj} + w)$ for $j = 1, 2, \ldots$

Step 3: If $k < (n-1)$, set $k = k + 1$ and return to 2.

Step 4: All nodes have been labelled, except the destination node. Find the label for the destination node, i.e., the minimum in-coming weight, let this minimum be *w*. The shortest distance from the origin to the destination node is *w* at the destination node '*n*' and path can be traced back by passing through the links with modified weights as zero or trace the first element of the label.

2.3.1 Numerical illustrations

2.3.1.1 Illustration 2.1(a): Minimum path

Let us reconsider the (8 nodes and 13 links) network taken from Munapo, Jones, and Kumar (2008), where they obtained the longest path, and here we will find the shortest path joining nodes 1 and 8. The network is shown in Figure 2.3 and corresponding link weights are displayed on the figure as well as they are given in Table 2.1.

After the initial label at node 1, node 2 will be labelled (3,1) and it will alter the link weights for the incoming and outgoing links associated with node 2. These links are: (1,2), (2,4), and (2,6). These changes are shown in Table 2.2. The label at node 2 is (1,3).

From Table 2.2, we obtain Table 2.3, where node 3 is labelled and link weights will be modified for the links (1,3), (3,4) and (3,7) as shown in Table 2.3.

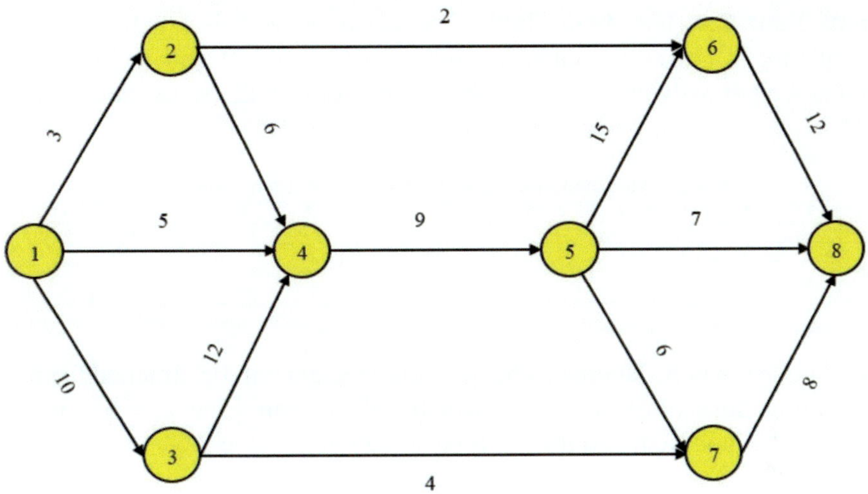

Figure 2.3 Directed network.

Table 2.1 Link and link weights from Figure 2.2.

S No	1	2	3	4	5	6	7	8	9	10	11	12	13
Link	1,2	1,3	1,4	2,4	2,6	3,4	3,7	4,5	5,6	5,7	5,8	6,8	7,8
Weight	3	10	5	6	2	12	4	9	15	6	7	12	8

Table 2.2 Modified link weights of various links after the label on node 2.

S No	1	2	3	4	5	6	7	8	9	10	11	12	13
Link	1,2	1,3	1,4	2,4	2,6	3,4	3,7	4,5	5,6	5,7	5,8	6,8	7,8
Weight LN 2(1,3)	3 − 3 = 0	10	5	6 + 3 = 9	2 + 3 = 5	12	4	9	15	6	7	12	8

Table 2.3 Modified link weights after the label on node 3.

S No	1	2	3	4	5	6	7	8	9	10	11	12	13
Link	1,2	1,3	1,4	2,4	2,6	3,4	3,7	4,5	5,6	5,7	5,8	6,8	7,8
Weight LN 3(1,10)	0	10 − 10 = 0	5	9	5	12 + 10 = 22	4 + 10 = 14	9	15	6	7	12	8

From Table 2.3, we obtain Table 2.4 where node 4 is labelled. The incoming links are (1,4), (2,4) and (3,4). The minimum incoming weight is = min{5,9, 22} = 5. It will change link weights of links (1,4), (2,4), (3,4), and the out-going link (4,5). The label on node 4 will be (1,5).

Table 2.4 Modified link weights from Table 2.3.

S No	1	2	3	4	5	6	7	8	9	10	11	12	13
Link	1,2	1,3	1,4	2,4	2,6	3,4	3,7	4,5	5,6	5,7	5,8	6,8	7,8
Weight, LN 4(1,5)	0	0	5 − 5 = 0	9 − 5 = 4	5	22 − 5 = 17	14	9 + 5 = 14	15	6	7	12	8

From Table 2.4, we obtain Table 2.5, where node 5 is labelled. The incoming link is (4,5) and outgoing links are (5,6), (5,7) and (5,8). The minimum weight will be = min $\{14\}$ the outcome will be as shown in Table 2.5.

Table 2.5 Modified link weights after the label on node 3.

S No	1	2	3	4	5	6	7	8	9	10	11	12	13
Link	1,2	1,3	1,4	2,4	2,6	3,4	3,7	4,5	5,6	5,7	5,8	6,8	7,8
Weight LN 5(4,5)	0	0	0	4	5	17	14	14 − 4 = 0	15 + 14 = 29	6 + 14 = 20	7 + 14 = 21	12	8

Next, node 6 will be labelled. The minimum weight will be obtained from two incoming links (2,6) and (5,6), which will be min $\{5{,}29\} = 5$. Link weights will be modified, and they will be as shown in Table 2.6.

Table 2.6 Modified weights obtained from Table 2.5.

S No	1	2	3	4	5	6	7	8	9	10	11	12	13
Link	1,2	1,3	1,4	2,4	2,6	3,4	3,7	4,5	5,6	5,7	5,8	6,8	7,8
Weight LN 6(2,5)	0	0	0	4	5 − 5 = 0	17	14	0	29 − 5 = 24	20	21	12 + 5 = 17	8

Node 7 will be labelled. It will alter link weights of links (3,7), (5,7), and (7,8). The minimum weight will be $\{14, 20\} = 14$. It will give Table 2.7.

Table 2.7 Obtained from Table 2.6.

S No	1	2	3	4	5	6	7	8	9	10	11	12	13
Link	1,2	1,3	1,4	2,4	2,6	3,4	3,7	4,5	5,6	5,7	5,8	6,8	7,8
Weight LN 7	0	0	0	4	0	17	14 − 14 = 0	0	24	20 − 14 = 6	21	17	8 + 14 = 22

Finally, we obtain Table 2.8 as given below. The minimum incoming weight is obtained from the link weights of links $\{(5,8), (6,8), (7,8)\} = \{21, 17, 22\}$. Hence the minimum path will be 17 units long. Only link will change (6,8).

Table 2.8 Final table.

S No	1	2	3	4	5	6	7	8	9	10	11	12	13
Link	1,2	1,3	1,4	2,4	2,6	3,4	3,7	4,5	5,6	5,7	5,8	6,8	7,8
Weight LN 8	0	0	0	4	0	17	0	0	24	20	21 − 17 = 4	17−17 = 0	22 −17 = 5

These labels are summarized in Table 2.9, and path from node 1 to node K are also detailed.

Table 2.9 Various labels and corresponding paths.

Node *k*	Label at node *k*	Path from node 1 to *k*
1	Initial label (1,0)	1 −> 1
2	(1,3)	2 −> 1
3	(1,10)	3 −> 1
4	(1,5)	4 −> 1
5	(4,14)	5 −> 4 −> 1
6	(2, 5)	6 −> 2 −>1
7	(3,14)	7 −> 3 −> 1
8	(6, 17)	8 −> 6 −> 2 −> 1

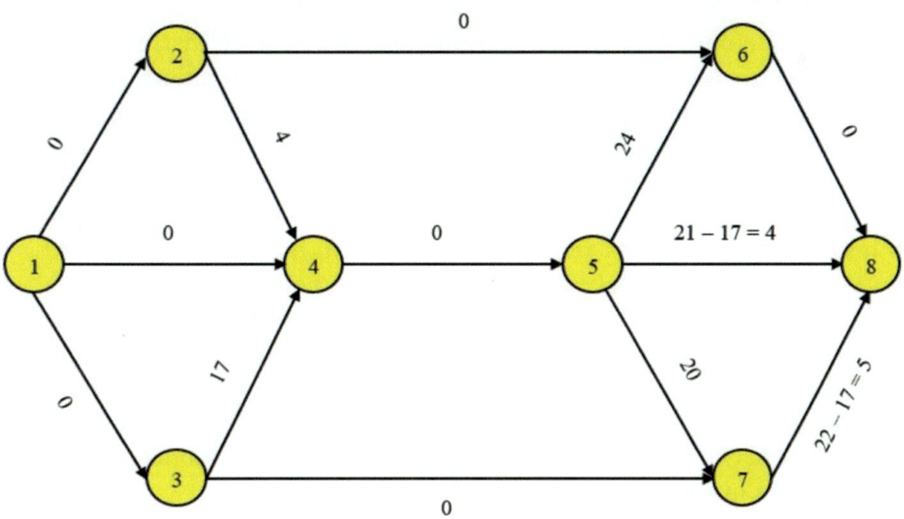

Figure 2.4 The network with modified link weights as given in Table 2.8.

Note that Figure 2.4 will be used later, when we discuss ordered optimum.

2.3.1.2 Illustration 2.2

Once again, one can also find the longest path and establish that cost associated with the longest path will be 58 units and the path will be:

$$1 \rightarrow 3 \rightarrow 4 \rightarrow 5 \rightarrow 6 \rightarrow 8.$$

Note the maximum problem can be easily analysed by considering initial data as a negative of the given value.

2.4 Shortest Route in a Non-directed Network

A non-directed network can't be analysed on lines of a directed network as conversion of a non-directed network to a directed network by replacing each link by two separate links in opposite directions is not feasible as the network will become unwieldy to deal with and nodes will not be able to maintain the topological order that we had in a directed network. Therefore, a new approach is required. However, the following properties hold for the non-directed network.

2.4.1 Nature of a label associated with a node in a non-directed network

The shortest path must start from the origin node, the shortest distance from origin to origin is a known distance, it is 0. Hence the origin node can be labelled as was done in the case of a directed network. Since topographical order of nodes is missing in a non-directed network, we need a label that can provide not only the total distance from the origin, but the label should also help to trace the path from the origin to that node. In other words, we need history associated with that label. Note that in the case of a directed network, a two-element label was enough but for a non-directed network a three-element label is required. It is given by (i, j, d), where i indicates the order, j indicates the previous node and d denotes the distance from the origin node to that node. Thus the last two elements in the three link label have similar interpretation as was the case for the directed network. Since nodes in a non-directed network are not in the topological order, an additional element takes care of the history associated with the path.

In this sense, the origin node of a non-directed will have a label (1,1,0), i.e., the label on the origin node will be: $1_{(1,1,0)}$, indicating distance is 0, we arrive at node 1 from node 1 and it is the first label. Similarly, labels on all other nodes will be three-element labels, giving the order of labelling at that node, the previous node from where we arrive at the current node and the total distance from the origin node to the current node.

2.4.2 Notation, definitions, and the concept of virtual directions

Some notation and definitions will facilitate the subsequent discussion. Define a few sets of nodes and links as follows:

L_0 = set of labelled nodes, which at the start, has only one element in this set, i.e., the origin node = $\{1\}$

L_1 = set of links connected to the labelled nodes. Initially all links (1 −> *j*) will belong to this set, as origin node '1' is the only labelled node.
L_2 = set of links that will never participate in the shortest path. Initially we have no knowledge about these links; hence it will be a null set $= \{\emptyset\}$.
$L_{3(1,k)}$ = set of links that form the shortest path between nodes '1' and node *k*. Initially the origin node is the only labelled node, hence, when $k = 1$, $L_{3(1,k)} = \{(1,1)\}$, i.e., no movement has been observed, or simply we moved from node '1' 'to' '1'. The path is *(1 −>1)*.

Since the first link on the shortest path will be a link starting from the origin node, one may assume the links from the origin as directed *(1 −> j)*, where *(1 −> j)* are directly connected links with the origin node. This is called the concept of **virtual directions**, as discussed by Kumar et al. (2013).

Although the links are non-directed, however, all links emanating from the origin can be assumed as directed links. The directions are from the origin node to other nodes going away from the origin node. In theory, one can return to the origin node '1' from any other node but the path from '1' to *j* will only increase its total distance. Thus, returning to a labelled node will increase the total distance and all links emanating from a labelled node have a virtual direction. All non-directed links, between two labelled nodes will only increase the total distance. These links will belong to the set L_2, i.e., the links will never be a part of the shortest route.

2.4.3 Link weight modification using the concept of virtual directions

Find the minimum weight associated with links that belongs to the set $\{L_1\}$. Let the corresponding node associated with the minimum weight be denoted by node j and the minimum weight be denoted by $w_{(1j)}$. Now label node j by $j_{(2\ ,1,\ w_{(1,j)})}$. The number 2 in the label indicates that it is the second label after the initial label at the origin node 1. The minimum distance from node 1 to j is $w_{(1j)}$ and the shortest path from the origin to node j is formed by the link $(1, j)$.

The link weights will be modified as follows:

New weight of the link $(1, j)$ will be $= w_{(1j)} - w_{(1j)} = 0$.
New weights associated with links (j, k) will be given by $w_{(j,k)} + w_{(1j)}$. The weight of the link $(1, j)$ in the direction $(j, 1)$ will be $= w_{(1j)} + w_{(1j)} = 2w_{(1j)}$. Since we will never return to a labelled node, there is no point about weight of the return link.

New weights of the remaining links will remain unchanged. And one must upgrade the sets L_1, L_2, and L_3.

Since nodes 1 and j are labelled, once again one can assign virtual directions of all links from the labelled nodes to unlabelled nodes. To find the min weight of all links in the set L_1, which will have all links joining nodes 1 and j to unlabelled nodes. Note a new node will be labelled each time, the process will terminate in n or fewer number of steps. In the case of the directed network, convergence occurred in 'n' number of iterations.

Thus, once again, the link weights have been altered to make link weight of at least one link as zero by identifying implied directions as we did in the previous iteration. For details, see Kumar et al. (2013).

2.5 Shortest Path in the Non-directed Network by Labelling Approach

The algorithmic steps are similar but not the same. These steps are:

Step 1: Label the origin node 1 as (1,1,0) indicating that the minimum distance from the origin node to itself is 0, it is the first labelled node, and we arrived at node 1 from node 1. The three sets at this stage are:

$$L_1 = \{1\}\,,\ L_2 = \{\varphi\}\,,\ L_3 = \{1\,(1-1)\}\,.$$

Step 2: Find links that are connected to labelled nodes. Find the minimum weight associated with these links and update the link weights, by subtracting the minimum and adding the minimum to all outgoing links. Update the sets L_1, L_2, and L_3. If the destination node is not labelled, repeat Step 2, else go to Step 3.

Step 3: If the destination node is labelled, the shortest route has been identified, find the shortest route and the shortest distance.

2.5.1 Numerical illustration for the non-directed network: Illustration 2.3

Consider the non-directed network in Figure 2.3, link weights are given in Table 2.10.

Since initially only the origin node, i.e., node 1 can be labelled as indicated in Table 2.10. Therefore, we can assign virtual directions to links (1,2), (1,3), (1,4), i.e., links from node 1 to k, where $k = 2, 3, 4$. The minimum weight of these links from node 1 is $= \{5,\ 3,\ 1\} = 1$, which corresponds to node 3. Hence, this node will be the next to be labelled as shown in Table 2.11.

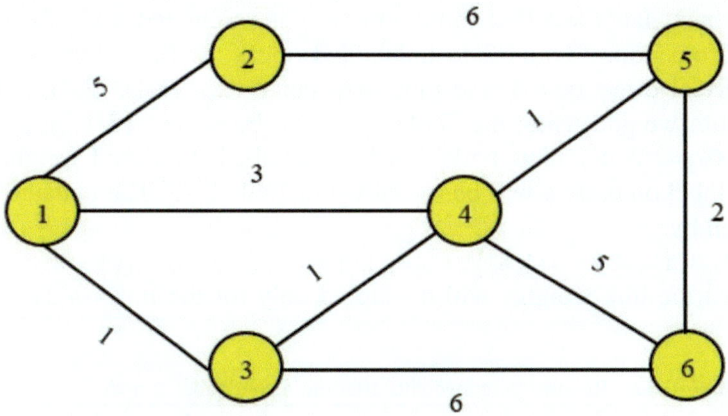

Figure 2.5 Non-directed network.

Table 2.10 Link weights for the network in Figure 2.5.

From/to	1	2	3	4	5	6	Label (i, j, d)
1	-	5_{12}	1_{13}	3_{14}	∞	∞	$1_{(1,1,0)}$.
2	5_{21}	-	∞	∞	6_{25}	∞	
3	1_{31}	∞	-	1_{34}	∞	6_{36}	
4	3_{41}	∞	1_{43}	-	1_{45}	5_{46}	
5	∞	6_{52}	∞	1_{54}	-	2_{56}	
6	∞	∞	6_{63}	5_{64}	2_{65}	-	

The modified weight for various links will be:

Link (1,3) $= 1 - 1 = 0$, link (3,1) $= 1 + 1 = 2$.
Link (3,4) $= 1 + 1 = 2$, and link (3,6) $= 6 + 1 = 7$. These modified weights are shown in Table 2.11.

Note the difference between the link weight modification that was carried out when we considered the directed network. Here we are considering only one link from node 1 to 'j'.

Note that nodes 1 and 3 are labelled nodes. The updated sets will be as follows: $L_1 = \{1, 3\}$, $L_2 = \{\varphi\}$, $L_3 = \{1(1-1), 3(1-3)\}$.

Table 2.11 Reconstructed weights for the network in Figure 2.3.

From/to	1	2	3	4	5	6	Label (i, j, d)
1	-	5_{12}	0_{13}	3_{14}	∞	∞	$1_{(1,1,0)}$.
2	5_{21}	-	∞	∞	6_{25}	∞	
3	2_{31}	∞	-	2_{34}	∞	7_{36}	$3_{(2,1,1)}$.
4	3_{41}	∞	1_{43}	-	1_{45}	5_{46}	
5	∞	6_{52}	∞	1_{54}	-	2_{56}	
6	∞	∞	6_{63}	5_{64}	2_{65}	-	

For the next label we find minimum weight associated with the links connected from nodes 1 and 3 to unlabelled nodes, as these links can be assigned virtual directions. These links will be: {(1,2), (1,4), (3,4), (3,6)}. The minimum weight, using the Table 2.11, will be = min {5(1,2), 3(1,4), 2(3,4), 7(3,6)}=$2(3,4)$. Thus node 4 will be labelled. Modified weights as well as the label on node 4 will be, as shown in Table 2.12. The updated sets will be given by:

$L_1 = \{1,\ 3, 4\}\ ,\ L_2 = \{(1,4)\}\ , L_3 = \{1\,(1-1)\ ,\ 3\,(1-3)\ , 4\,(4-3-1)\}\ .$

Note that the link weights will be altered only for the links (4,3), (4,1), (4,5), (4,6).

Table 2.12 Reconstructed weights after the second label at node 4.

From/to	**1**	**2**	**3**	**4**	**5**	**6**	**Label (i, j, d)**
1	-	5_{12}	0_{13}	$(3-2) = 1_{14}$	∞	∞	$1_{(1,1,0)}.$
2	5_{21}	-	∞	∞	6_{25}	∞	
3	2_{31}	∞	-	0_{34}	∞	7_{36}	$3_{(2,1,1)}.$
4	∞_{41}	∞	∞_{43}	-	3_{45}	7_{46}	$4_{(3,3,2)}$
5	∞	6_{52}	∞	1_{54}	-	2_{56}	
6	∞	∞	6_{63}	5_{64}	2_{65}	-	

The new set of links with virtual directions will be links joining the labelled nodes to unlabelled nodes, which are: {(1,2), (3,6), (4,5), (4,6)}. The corresponding minimum will be = $\min\left(5(1,2), 7(3,6), 3(4,5), 7(4,6)\right) = 3(4,5)$. Hence node 5 will be the next node that will be labelled. The reconstructed weights and the new label are displayed in Table 2.13. The updated values of the sets will be:

$$L_1 = \{1,\ 3, 4, 5\},\ L_2 = \{(1,4)\},$$
$$L_3 = \{1(1-1),\ 1(1-3), 2(4-3-1), 3(5-4-3-1)\}.$$

This updated information is given in Table 2.13.

Table 2.13 Updated information from Table 2.12.

From/to	**1**	**2**	**3**	**4**	**5**	**6**	**Label (i, j, d)**
1	-	5_{12}	0_{13}	3_{14}	∞	∞	$1_{(1,1,0)}.$
2	5_{21}	-	∞	∞	6_{25}	∞	
3	2_{31}	∞	-	0_{34}	∞	7_{36}	$3_{(2,1,1)}.$
4	∞_{41}	∞	∞_{43}	-	0_{45}	7_{46}	$4_{(3,3,2)}$
5	∞	6_{52}	∞	1_{54}	-	5_{56}	$5_{(4,4,3)}$
6	∞	∞	6_{63}	5_{64}	2_{65}	-	$6_{(5,4,5)}$

Since node 6 remains unlabelled, we once again repeat the iteration and find the set of links which join the labelled node to the unlabelled nodes. These links are given by: {(1,2), (3,6), (4,6), (5,6)} and the minimum will be = {5(1,2), 7(3,6), 7(4,6), 5(5,6)}=5 for two nodes, i.e., nodes 2 and 6. Since node 6 is the required destination node, we select that and conclude the minimum distance path as shown in Table 2.14. Updated sets will be:

$$L_1 = 1, 3, 4, 5, 6, L_2 = (1, 4), (1, 2),$$
$$L_3 = \{1(1-1), 3(1-3), 4(4-3-1), 5(5-4-3-1), 6(6-5-4-3-1)\}.$$

Table 2.14 The final table.

From/to	1	2	3	4	5	6	Label (i, j, d)
1	-	5_{12}	0_{13}	3_{14}	∞	∞	$1_{(1,1,0)}$.
2	5_{21}	-	∞	∞	6_{25}	∞	
3	2_{31}	∞	-	0_{34}	∞	7_{36}	$3_{(2,1,1)}$.
4	∞_{41}	∞	∞_{43}	-	0_{45}	7_{46}	$4_{(3,3,2)}$
5	∞	6_{52}	∞	1_{54}	-	0_{56}	$5_{(4,4,3)}$
6	∞	∞	∞	∞	∞	-	$6_{(5,4,5)}$

2.5.2 Illustration 2.4

Consider a non-directed network in Figure 2.6 and identify the shortest path and the associated weight by using the reconstruction approach discussed in this chapter.

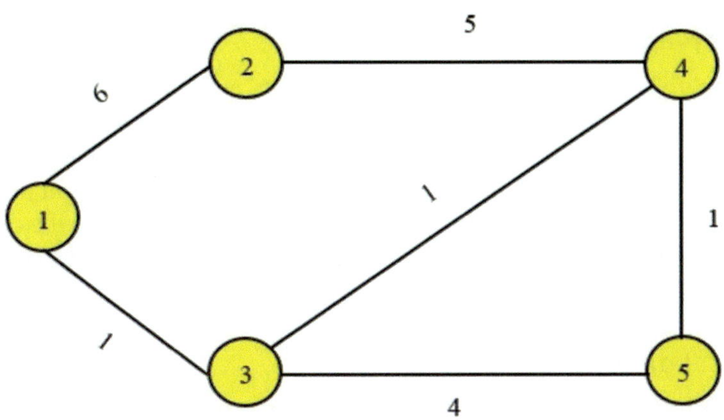

Figure 2.6 Network for the minimum and the maximum route.

This trivial example has only three alternative paths as given below: 12 ($1 \to 2 \to 4 \to 5$), 3 ($1 \to 3 \to 4 \to 5$) and 5 ($1 \to 3 \to 5$). Use the network reconstruction method and identify the minimum and the maximum optimal routes.

2.6 An Application of Virtual Directions

The concept of virtual directions has been applied to the problem of finding the maximum reliability route in a non-directed reliability graph by Kumar et al. (2016). In the case of a reliability network, links are associated with probability of communication without any distortion, and nodes are assumed perfect, i.e., the reliability associated with a node is 1. The following observations are immediate:

2.6.1 Observation 2.1

Two numbers 'p' and 'q' representing probabilities, i.e., the values of 'p' and 'q' are such that they will satisfy the inequality:

$$0 \leq p, q \leq 1. \tag{2.1}$$

In that case they will also satisfy the inequality in eqn (2.2)

$$0 \leq p.q \leq \min\left(p, q\right). \tag{2.2}$$

Proof

Let the min $(p, q) = p$. Note that $p - p.q = p.(1 - q) \geq 0$. Similarly, one can reiterate steps for the probability q.

2.6.2 Observation 2.2

Although the given network has non-directed links, a revisit to the origin or any other node will reduce the reliability compared to its present reliability, which is 1. Hence, like the routing in a non-directed network, once again all links (1,*j*) can be assigned virtual directions as ($1 \to j$) and the origin node 1 can be labelled as $1_{(1,\ 1,1)}$, meaning that the first label is assigned to the origin node, we reach to node 1 from node1, with node reliability 1 and therefore the reliability will remain 1.

2.6.3 Virtual directions theorem

Once a node has been labelled, this label creates a set of links that can be assigned virtual directions. These links are those connecting the labelled node to unlabelled nodes. The direction will be from the labelled node towards the unlabelled node.

Initially the origin node will be the only labelled node. The virtual directions will be from the origin node to directly connected unlabelled nodes. This will help developing a new label and iteratively, one can label the terminal node to find the maximum reliability in the given reliability network *G*(*n*,*m*), the reliability can be evaluated in at most '*n*' iterations.

2.6.4 Numerical illustration 2.5

Consider the 8 node 13 link probability network given in Figure 2.7. The probability values are shown on each link.

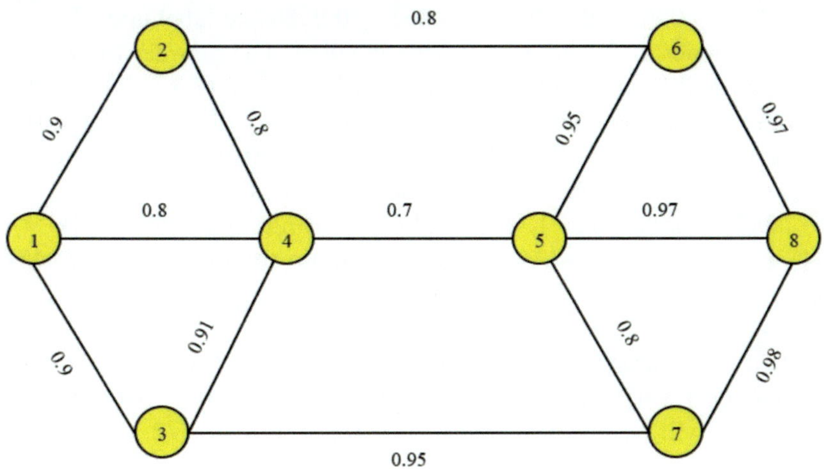

Figure 2.7 Probability network.

The network in Figure 2.7 is a non-directed network, the probability from i to j will be the same as from j to i. It is shown in Table 2.15.

Table 2.15 Link probabilities from Figure 2.7.

$i \backslash j$	1	2	3	4	5	6	7	8
1	1	0.9	0.9	0.8	0	0	0	0
2	0.9	1	0	0.8	0	0.8	0	0
3	0.9	0	1	0.91	0	0	0.95	0
4	0.8	0.8	0.91	1	0.7	0	0	0
5	0	0	0	0.7	1	0.95	0.8	0.97
6	0	0.8	0	0	0.95	1	0	0.97
7	0	0	0.95	0	0.8	0	1	0.98
8	0	0	0	0	0.97	0.97	0.98	1

Calculations:

1. The initial label at node 1 will be 1(1,1,1), indicating that it is the first label, we reach node 1 from node 1, which is perfect, i.e., the probability is 1.
2. After this initial label, the links directly connected to node 1 are: (1,2), (1,3), and (1,4). These links get the virtual directions from node 1 to k, k = 2, 3, 4. The maximum reliability is {0.9, 0.9, 0.8} = 0.9 for two nodes: 2 and 3. Choose arbitrarily, node 3. The label at node 3 will be (2, 1, 0.9) indicating that it is the second label, we arrive at this node from node 1 and the maximum probability is 0.9.
3. Note the labelled nodes are 1 and 3, feasible links for the virtual direction will be {(1,2), (1,4), 3,4), (3,7)} = Max {1 × 0.9 = 0.9, 1 × 0.8 = 0.8, 0.9 × 0.91 = 0.86, 0.9 × 0.95 = 0.86} = 0.9. Hence label node 2.
4. Note that nodes 1, 2, and 3 have been labelled. The new set for virtual directions will be made of: {(1,4), (2,4), (2,6), (3,4), (3,7)}. The corresponding reliabilities will be given by: max {1 × 0.8 = 0.8, 0.9 × 0.8 = 0.72, 0.9 × 0.95 = 0.86, 0.9 × 0.95 = 0.86} = 0.86 corresponding to nodes 6 or 7. We arbitrarily select node 6, which is labelled next.
5. The labelled nodes are {1, 2, 3, and 6}. Feasible links will be: {(1,4), (2,4), (3,4), (3.7), (6,5), (6,8)}. The maximum reliability will be Max {0.8, 0.72, 0.82, 0.86, 0.66, 0.83} = 0.86 corresponding to node 7, which is the next node to be labelled.
6. Labelled nodes are (1, 2, 3, 6, and 7). Feasible links with virtual directions will be {(1,4), (2,4), (3,4), (6,5), (6,8), (7,5), (7,8)}. Corresponding reliabilities will be given by: Max {0.8, 0.9 × 0.8 = 0.72, 0.9 × 0.91 = 0.82, 0.86 × 0.95 = 0.731, 0.86 × 0.97 = 0.83, 0.86×0.8 = 0.69, 0.86×0.98 = 0.84} = 0.84 node 8 reached from node 7.

Table 2.16 Analysis summary of the network in Figure 2.5.

Serial no	Label	Path	Remarks
1	1(1,1,1)	1 →1	Item 1
2	3(2,1,0.9)	3 →1	Item 2
3	2(3,1,0.9)	2 →1	Item 3
4	6(4,2,0.86)	6 → 2 →1	Item 4
5	7(5,3,0.86)	7 → 3 →1	Item 5
6	**8(6,7,0.84)**	**8 → 7 → 3 → 1**	Item 6
7*	4(7,3,0.82)	7 → 4 → 3 → 1	Item 7
8*	5(8,7,0.814)	8 →7 → 3 → 1	Item 7

7. Maximum reliability path has been identified. However, if we continue to label the remaining nodes, they will be as given in Table 2.16. Detailed calculations are not given.

2.7 *K*th Shortest Route

The $K\text{th}(K \geq 2)$, shortest route in a network is defined as the shortest route after excluding ($K - 1$) number of shortest routes. There is no well-defined theory for determination of order optimal solutions to optimization problems. Here we will consider a shortest route and an assignment problem to find the second-best solution, using a concept of creating an alternative optimal solution for a given optimal situation. Many suggestions have been proposed in the past, but they can be computationally demanding. Optimization problems arise in various real-life situations, and in general determination of the best or the optimal solution can be identified as all existing methods have a defined condition for the optimality of the solution, and when those conditions are satisfied, one can easily conclude that the optimal solution has been obtained. However, existence of the optimality conditions for the *K*th best solution, $K \geq 2$ have not been defined, rather the $K\text{th}$ optimal is defined as the best, excluding ($K - 1$) number of best solutions. A need for the *K*th best solution arises in many different situations, for example see, Kumar and Wagner (1979), where the $K\text{th}$ best is required for a linear program for analysing the extreme point mathematical programming problem. Al-Hasani et al. (2020) used the ordered optimal solution to find the set of non-dominated points in a multi-objective assignment problem. Huynh, Kumar and Connell (1987) applied the random search method for integer programming model, Huynh, Kumar and Connell (1987) developed the random search method for the mixed integer programming model, and later Huynh and Kumar (1990) applied these ideas to solve the extreme point mathematical programming problem. For more details about the random search method, see Chapter 6, Munapo and kumar (2022).

From the above, the $K\text{th}$ best solution is required for different reasons in different situations. In response to this, many researchers have developed ways to identify the ordered optimal solution. For example, Murthy (1968) proposed a method for ranking all the assignments in order of increasing costs. In an $n \times n$ assignment problem, Murthy (1968) obtained the second best by solving *n* assignment problems by assigning high cost to one at a time of the optimal assignments, and the best of all these *n* solutions as the second best solution for that problem. Munapo and kumar (2022) resolved Murthy

(1968) example by making an assignment at the least cost element and identified zero cost solution in the remaining $(n-1) \times (n-1)$ assignment problem. They established the same solution as was obtained by Murthy (1968).

Here in this chapter, we introduce yet another approach to identify the Kth best solution and present a couple of numerical illustrations.

The main idea is to increase the length of the current optimal solution by certain amount to establish two or more alternative optimal solutions. One of these alternative solutions will have to be the existing optimal solution and the other is a newly created alternative optimal solution. Since the value of the optimal solution is increased in a controlled way, these two alternative optimal solutions have an interesting interpretation of the best and the second best solution with respect to the original cost associated information. We illustrate the idea by taking the shortest route problem in a network and an assignment problem to find the best and the second best solutions for these two problems.

2.7.1 Second best route for the shortest route problem

2.7.1.1 Numerical illustration 2.6

Let us apply the idea of developing alternative solutions to the network routing problem considered in Figure 2.8, which was transformed into the Figure 2.9, while finding the optimal solution.

For convenience, both figures are reproduced below, as we shall be extracting some important information for finding the second best solution. The optimal solution was identified from Figure 2.9 as total cost 0, which from Figure 2.8 was 17 units, and the path was $1 \to 2 \to 6 \to 8$.

This path was obtained from Figure 2.9.

From Figure 2.4, we see that the least non-negative value > 0 in Figure 2.9 is 4 with respect to two links, i.e., links (2,4) and (5,8). Therefore, the next shortest path compared to the current optimal path will be at least 4 units longer.

The main idea is to increase the length of the current optimal solution by 4 units and if we can establish two alternative optimal paths of total length 21 units, after increasing the length of the current optimal path by 4 units, then one of them, the previous optimal path and the other is a new path that has a length equal to the increased length. The previous path will be the best solution and the other path will be the second best solution. Therefore, from these two alternative paths, we establish the optimal and the second best optimal solutions to the shortest route problem.

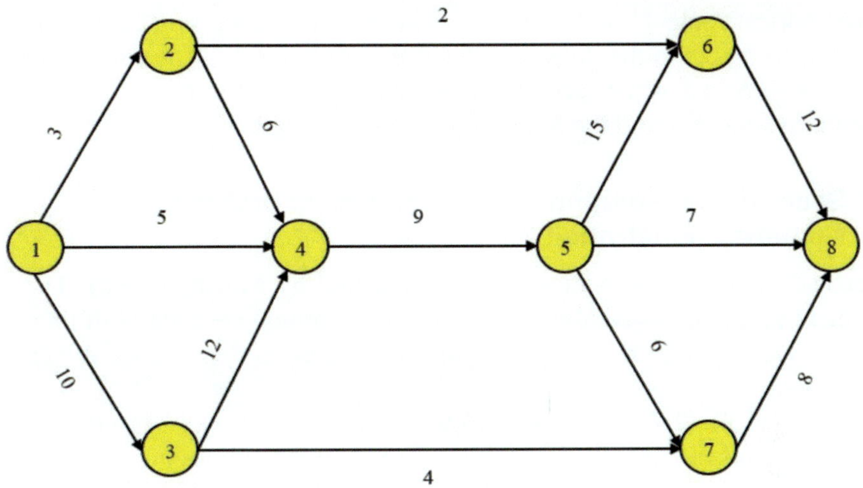

Figure 2.8 Directed network.

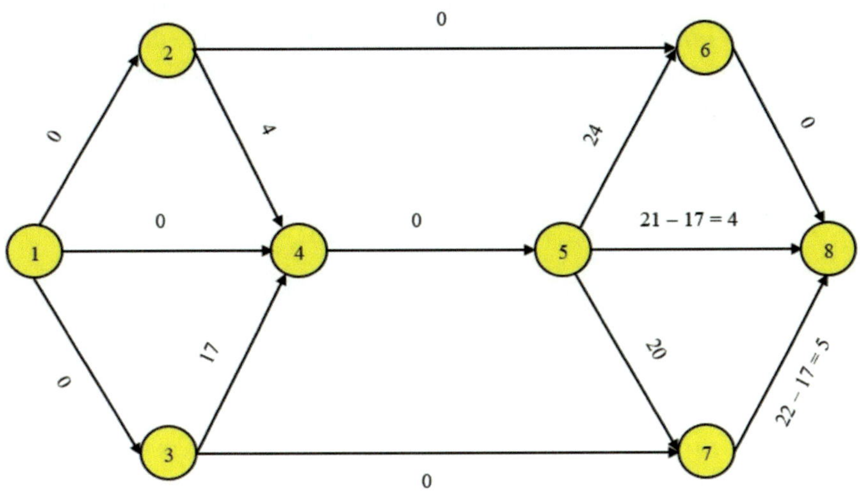

Figure 2.9 Reproduced for convenience of presentation.

Use of the link (2,4) does not create another path joining nodes 1 and 8, however, the link (5,8) does develop a path $1 \rightarrow 4 \rightarrow 5 \rightarrow 8$, with total length equal to 21, becoming the second shortest path for the network in Figure 2.2. If we increase length of any link (1,2), (2,6), or (6,8) by 4 units, we end up getting two alternative solutions. Therefore, we conclude that the:

The shortest route $1 \rightarrow 2 \rightarrow 6 \rightarrow 8$ with modified length of 21 units is shortest with respect to the given link-lengths, cost being 17. The alternative path $1 \rightarrow 4 \rightarrow 5 \rightarrow 8$ is the second best and its length is 21 units.

Similarly, one can find the 3rd best path and its length.

2.7.2 Second best solution for an assignment problem: Numerical illustration 2.7

We reconsider the assignment problem considered by Murthy (1968). The assignment problem is given in Table 2.17. The optimal solution with total cost 0 is given by the following assignments, which is denoted by A (1), given below:

A (1) = {(1,9), (2,7), (3,3), (4,8), (5,6), (6,4), (7,10), (8,1), (9,5), (10,2)}. These allocations in Table 2.17 are shown in red. Vertical lines are drawn to cover zero entries in columns 9, 3, 4, 1, 5, 2, and 7. Rows are covered by three horizontal lines 4, 5, and 7. For the second-best assignment, Murthy solved 10 more problems by setting one assignment from the optimal solution at a time equal to ∞ . For example, if we set the link (1,9) = ∞, the problem can be solved again. A summary of the solutions that Murthy obtained are given below and denoted by eqn (2.3):

$$M_1 = \{(\overline{(1,9)}\} = 10,$$

$$M_2 = \{(1,9),\ \overline{(2,7)}\} = 14$$

$$M_3 = \{(1,9),\ (2,7),\ \overline{(3,3)}\} = \text{Cost } 14$$

$$M_4 = \{(1,9),\ (2,7),\ (3,3),\ (\overline{(4,8)}\} = \text{Cost } 1$$

$$M_5 = \{(1,9),\ (2,7),\ (3,3),\ (4,8),\ \overline{(5,6)}\} = \text{Cost } 12$$

$$M_6 = \{(1,9),\ (2,7),\ (3,3),\ (4,8),\ (5,6),\ \overline{(6,4)}\} = \text{Cost } 53$$

$$M_7 = \{(1,9),\ (2,7),\ (3,3),\ (4,8),\ (5,6),\ (6,4),\ \overline{(7,10)}\} = \text{Cost } 45$$

$$M_8 = \{(1,9),\ (2,7),\ (3,3),\ (4,8),\ (5,6),\ (6,4),\ (7,10),\ \overline{(8,1)}\} = \text{Cost } 47$$

$$M_9 = \{(1,9),\ (2,7),\ (3,3),\ (4,8),\ (5,6),\ (6,4),\ (7,10),\ (8,1)\ \overline{(9,5)}\} = \text{Cost } 56. \tag{2.3}$$

From the above, he concluded that the second-best solution corresponds to the string M_4. This string of assignments is given by:

A (2) = {(1,9), (2,7), (3,3), (4,2), (5,6), (6,4), (7,8), (8,1), (9,5), (10,10)} with total cost 1.

Table 2.17 Assignment cost values for the Murthy's Illustration.

R\C	1	2	3	4	5	6	7	8	9 V↓ 1	10
1→	7	51	52	87	38	60	74	66	0	20
2	50	12	0	64	8	53	0	46	76	42
3	27	77	0	18	22	48	44	13	0	57
4	62	0	3	8	5	6	14	0	26	39
5	0	97	0	5	13	0	41	31	62	48
6	79	68	0	0	15	12	17	47	35	43
7	76	99	48	27	34	0	0	0	28	0
8	0	20	9	27	46	15	84	19	3	24
9	56	10	45	39	0	93	67	79	19	24
10	27	0	39	53	46	24	69	46	23	1

Now we apply the concept of creation of an alternative. From Table 2.17, the minimum non-zero element is 1 in the cell (10,10). In the column 10 and row 10, we have allocation, the optimal solution denoted by the string A (1) has allocations in the (7,10) and (10,2). To create an alternative, cost in row seven is increased by one, therefore, the relative merit will not change. New values in row 7 are shown in Table 2.18. Similarly, new values in column 2 will be increased by 1, which is shown in Table 2.19.

Table 2.18 Altered values in row 7 ($C_{(7,j)}$, $j = 1, 2, \ldots, 10$).

Row 7	76 + 1	99 + 1	48 + 1	27 + 1	34 + 1	0 + 1	0 + 1	0 + 1	28 + 1	0 + 1

Table 2.19 Altered cost values in column 2 $C^T_{(i,2)}$, $i = 1, 2, \ldots, 10$.

Col 2	51 + 1	12 + 1	77 + 1	0 + 1	97 + 1	68 + 1	99 + 1	20 + 1	10 + 1	0 + 1

These changes will result in Table 2.20, which is essentially the same as Table 2.17, except for the row 7 and the column 2.

Table 2.20 Modified cost values.

R\C	1	2	3	4	5	6	7	8	9 V	10
1→	7	52	52	87	38	60	74	66	0	20
2	50	13	0	64	8	53	0	46	76	42
3	27	78	0	18	22	48	44	13	0	57
4	62	1	3	8	5	6	14	0	26	39
5	0	98	0	5	13	0	41	31	62	48
6	79	69	0	0	15	12	17	47	35	43
7	77	101	49	28	35	1	1	1	29	1
8	0	21	9	27	46	15	84	19	3	24
9	56	11	45	39	0	93	67	79	19	24
10	27	1	39	53	46	24	69	46	23	1

Note that rows 7 and 10 have no zero entry, when minimum is subtracted, and we get Table 2.21.

Table 2.21 A zero in each row and column.

R\C	1	2	3	4	5	6	7	8	9 V	10
1→	7	52	52	87	38	60	74	66	0	20
2	50	13	0	64	8	53	0	46	76	42
3	27	78	0	18	22	48	44	13	0	57
4	62	1	3	8	5	6	14	0	26	39
5	0	98	0	5	13	0	41	31	62	48
6	79	69	0	0	15	12	17	47	35	43
7	76	100	48	27	34	0	0	0	28	0
8	0	21	9	27	46	15	84	19	3	24
9	56	11	45	39	0	93	67	79	19	24
10	26	0	38	52	45	23	68	45	22	0

From Table 2.21, we obtain two alternative solutions given below, where the alternative links are in red.

First solution: {(1,9), (2,7), (3,3), (4,2), (5,6), (6,4), (7,8), (8,1), (9.5), (10,10)}

Second solution: {(1,9), (2,7), (3,3), (4,8), (5,6), (6,4), (7,10), (8,1), (9,5), (10,2)}.

With respect to the data information in Table 2.17, first solution is the second best with total cost 1. The second solution is the optimal solutions with total cost 0 as were obtained by Murthy (1968).

2.8 Concluding Remarks

In this chapter, a network reconstruction approach for solving minimum and maximum path in a directed and non-directed network has been discussed. This approach for the directed network was developed by Munapo, Jones and Kumar (2008) and for the non-directed network by Kumar et al. (2013). For the non-directed network, we have discussed a new concept of virtual directions; see Kumar et al. (2013, 2016).

The maximum or the minimum path in a directed network can be obtained in '*n*' iterations. However, the maximum or the minimum path in a non-directed network can be evaluated in at most *n* iterations, a problem that was considered computationally difficult by the dynamic programming approach, can get the result in fewer number of iterations than for the case of a directed network. Hence its computational complexity is better than for the case of directed network. However, in both cases, i.e., for the directed and the non-directed network converges is guaranteed in at most '*n*' iterations.

The concept of virtual directions was developed and applied to solve a probability network problem for identification of the maximum reliability route and the corresponding reliability associated with that path.

We have introduced a new concept of developing an alternative optimum by increasing the cost associated with the optimum route, and thus creating an alternative optimum with one of these solutions as the existing answer. The concept of developing an alternative is also playing a role in determination of an ordered optimum. This idea needs further research and refinements. Here it is illustrated for determination of the second best and in a similar way, it can be extended for the third best solution.

More applications of various concepts developed in this chapter are future challenges for researchers and practitioners. The concept of link weight modification is an optimization tool, yet its execution is an art, which depends on our own ability to exploit the given situation to develop a suitable process. It is hoped that researchers may like to expand these ideas further and explore the concept of link weight modification to develop more optimization tools for solving other network optimization problems.

References

A. Al-Hasani, M. Al-Rabeeah, S. Kumar, and A. Eberhard, A rank-based solution method to determine non-dominated points for a multi-objective assignment problem, International Journal of Mathematical, Engineering and Management Sciences, Vol. 5, No 6, 1249-1269, 2020.

R. E. Bellman, Dynamic Programming, Princeton University Press, 1957.

R. E. Bellman, On an optimal routing problem, Q Appl. Maths., pp 87-90, 1958.

R. E. Bellman, and S. E. Dreyfus, Applied Dynamic Programming, Volume 2050 in the series Princeton Legacy Library, 2015 (Original book was published in 1962) https://doi.org/10.1515/9781400874651

G. B. Dantzig, On the shortest route through a network, Management Sci., 6(2), 187-90, 1960.

E. W. Dijkstra, A note on two problems in connections with graphs, Numerical Mathematics, 1: 269-271, 1959.

N. A. J. Hastings, Dynamic Programming with Management Applications, Amazone, 1973.

H. N. Huynh, S. Kumar, and H. Connell, Random search method for integer programming, Proc. of the International Conference on optimization btechniques, Eds. Teo, Paul, Chew and Wong, pp 601-610, 1987, University of Singapore.

H. N. Huynh, H. Connell, and S. Kumar, Random search Method for the mixed integer programming model, ASOR 1987 Proc of the OR Conference, Ed. S. Kumar, pp 109-118, 1987.

H. N. Huynh, and S. Kumar, A random search method for Extreme Point Mathematical Programming Problem, Asia Pacific Journal of Operations Research, Vol. 7, No 1, pp 30-46, 1990.

R. Kalaba, On communication network problems, A Chapter in Combinatorial analysis, Proc. of Symposium of Appl Math 10, 1960.

S. Kumar, and D. Wagner, some algorithms for solving extreme point mathematical programming problems, New Zealand Operations Research Journal, Vol. 7, No. 2, pp 127-149, 1979.

S. Kumar, E. Munapo, O. Ncube, C. Sigauke, and P. Nyamugure, A minimum weight labeling method for determination of a shortest route in a non-directed network, Int. J of Syst. Assur Eng. and Management. Vol 4, No. 1, pp 13-18, 2013.

S. Kumar, E. Munapo, P. Nyamugure, and T. Tawanda, Mathematics of OR: Significance and Application of virtual directions in reducing computations complexity in network optimization, Chapter 3, Emerging trends in Applied Research, Vol 3, (Editor) IS Chauhan, Integrated Publications, 2022, Delhi, India.

S. Kumar, E. Munapo, M. Lesaoana, and P. Nyamugure, Identification and application of virtual directions in a non-directed network: An labelling method for determination of maximum reliability and the route, Communications in Dependability and Quality Management - An International Journal 19 (1), pp 85-95, 2016.

S. Kumar, E. Munapo, M. Lesaoana, and P. Nyamugure, Some innovations in OR Methodology: Linear Optimization, Lambert Academic Publication, 2018, ISBN 987-613-7-38007-9

E. Munapo, B. C. Jones, and S. Kumar, S., A minimum incoming weight label method and its applications in CPM Networks, ORiON, Vol. 24, No 1, pp 37-48, 2008

E. Munapo, and S. Kumar, Linear Integer Programming: Theory, Applications and Recent Developments, De Gruyter Series on Applications of Mathematics in Engineering and Information Sciences, Edited by Mange Ram, 2022, ISBN 978-3-11-070292-7, https://doi.org/10.1515/97831107 03023-202

K. G. Murthy, An algorithm for ranking all assignments in order of increasing cost, Operations Research, 16, 3, pp 682-687, 1968.

R. M. Pert, R. M. Randolf, T. E. Battlett, The shortest route problem, Operations Res 8, pp 866-868, 1960.

M. Pollack, W. Wiebenson, Solution of the Shortest-Route Problem–A Review Op. Res. **8** (2): 224–230, 1960.

J. P. Saksena, Dynamic programming applications, Siddharth Publications, Delhi, ISBN: 81-7220-184-2, 2007.

J. P. Saksena, and S. Kumar, The routing problem through K specified nodes, Opns Res 14, 909-913, 1966.

3

Network Reconstruction Approach to a Minimum Spanning Tree under an Index Restriction

Abstract

In this chapter the minimum spanning tree problem has been considered under the node index restriction by using the network reconstruction approach. The spanning tree under the index restriction has the potential to analyse NP hard combinatorial minimum travelling salesman tour problem, which has many industrial applications. In Chapter 5, the travelling salesman tour has been discussed using the minimum spanning tree under an index restriction.

Keywords: Non-directed network, tree of a network, minimum spanning tree, minimum spanning tree under the index restriction, network reconstruction approach, index balancing theorem.

Notation

1. $G(n, m)$ represents a non-directed connected network, where 'n' represents the set of nodes and 'm' represents the set of links joining various pairs of nodes. The number of elements in these two sets is denoted by 'N' and 'M', respectively.
2. (i, j) represents a typical link in the non-directed network joining the nodes i and j, $(i, j = 1, 2, \ldots, N$ and links $(i, j) = 1, 2, \ldots, M)$.
3. Let n_i represent the node index at the node i of the network $G(n, m)$, and let t_i represent the node index for the minimum spanning tree $T(N, (N - 1))$ of the given non-directed network $G(n, m)$.
4. L(MST) is the length of the minimum spanning tree of the network $G(n, m)$.

3.1 Introduction

In this chapter, we reconsider a non-directed connected network $G\,(n,\ m)$, where the set '*n*' denotes a set of nodes and '*m*' denotes a set of all weighted links joining various pairs of nodes in the network. Let the number of elements in these two sets be denoted by '*N*' and '*M*', respectively. A spanning tree of the non-directed network $G\,(n,\ m)$ is a connected subnetwork that includes all '*N*' nodes and only '*(N – 1)*' links from the set of links in the set '*m*'. This subnetwork is a tree, i.e., a network free of loops and circuits and remains connected. The tree of the given network becomes a minimum spanning tree when the sum of all selected links is minimum. It is a classical network problem, which has many industrial applications. For example, the minimum spanning tree has applications in minimizing construction costs associated with power networks, wiring connections, automatic speech recognition, etc. People often use algorithms that gradually build a spanning tree. Given the network and the pair-wise weights between nodes, the minimum spanning tree can be obtained iteratively by the greedy approach in (*N* – 1) iterations, see Anupam, (2015), Garg and Kumar, (1968). The process starts arbitrarily from any node and connects that node to the nearest node, forming a minimum spanning tree of the two nodes. Next, we find one more nodes that join any one of the previous two nodes by the nearest distance, forming a minimum spanning tree of three nodes. In other words, the selected edge should not form a loop with any existing selected nodes. Like this, in (*N* – 1) iterations, one can connect all *N* nodes forming the minimum spanning tree.

For a history of the minimum spanning tree (MST), see Graham and Pavol, (1985). Additional constraints and time windows for the spanning tree have been considered by Nemahauser and Wolsey (1989), Padberg and Rinaldi, (1991), Akgun and Tansel, (2010). Matai et al. (2010) have presented a review of methods and applications of the minimum spanning tree. The MST has also been used in DNA molecule computations, see Peppino et al. (2013) and Zhaocai et al. (2013). An evolutionary algorithm was developed by Li et al. (2013) for the minimum spanning tree. MST has been applied in assessing asset returns; see Resovsky et al. (2014). In graph theory a variation of MST was proposed by Xuong, discussed in (2009). Let the spanning tree of the graph $G\,(n,\ m)$ be denoted by *T*(*N*, *N* – 1), then the graph T has a property that in the remaining graph $\{G\,(n,\ m)$ – *T(N, N – 1)*$\}$ is such that the number of connected components with odd number of edges is minimum.

The minimum spanning tree under node index restriction was developed by Munapo et al. (2016) and its applications to the minimum travelling salesman tour were discussed by Kumar et al. (2016, 2018, 2019). For some other approaches dealing with the travelling salesman problem can be seen in Mitchell (2001), Nadef (2002), Cowen (2011), and Munapo & Kumar (2022).

Section 3.2 briefly presents the greedy approach for determination of the minimum spanning tree. Sections 3.3 and 3.4 deal with the determination of the minimum spanning tree under the index restriction. Numerical illustrations have been discussed in Section 3.5. Some concluding remarks are presented in Section 3.6.

3.2 The Minimum Spanning Tree Algorithm: Greedy Approach

The greedy approach is well documented in the OR literature, see Winston (2004). However, we briefly state the steps, which are as follows:

Step 1: Sort all the edges in the set $\{m\}$ from low weight to high of the given network $G(n, m)$. Select the lowest weight edge and include it as a part of the minimum spanning tree. Set $k = 1$.

Step 2: Set, $k = k + 1$ and take the kth edge with the lowest weight and add it to the spanning tree. If the edge to be added creates a cycle, then reject the edge and select the next one. If $k < (N - 1)$, return to Step 2, else go to Step 3.

Step 3: The minimum spanning tree has been identified and it is comprised of the selected $(N - 1)$ edges.

3.3 Statement of the Problem and Necessary Mathematical Support

3.3.1 Statement of the problem

For a given network $G(n, m)$, the minimum spanning tree can be obtained by $(N - 1)$ greedy iterations as discussed in Section 3.2. Let n_i represent the node index at the node i of the given graph $G(n, m)$ and let t_i represent the node index for the minimum spanning tree $T(N, (N - 1))$ of the non-directed network $G(n, m)$. Note that:

$$1 \leq t_i \leq n_i \leq (N-1) \text{ for } \forall i,\ i = 1, 2, \ldots, N. \tag{3.1}$$

Therefore, the minimum value of t_i and n_i for any node is 1 and the maximum value of t_i and n_i is (N – 1). It can happen when all nodes in the minimum spanning tree are connected to the same node. Under the node index restriction, the modified tree will have:

$$1 \leq t_i \leq 2 \text{ for } \forall i,\ i = 1,\ 2,\ \ldots,\ N. \tag{3.2}$$

Therefore, the minimum spanning tree obtained by the greedy approach needs to have some modifications to satisfy the condition (3.2).

For carrying out the required modification, we need some more definitions given in Section 3.3.2.

3.3.2 Terms and definitions

Index of a node, represented by t_i and n_i, $i = 1,\ 2,\ \ldots,\ N$; is a number representing the number of arcs emanating from the node i. Since the total number of selected arcs in the MST will be equal to $(N-1)$, the total index value of all the selected arcs will be $2\,(N-1)$. In an extreme case, the index number can be distributed to various nodes such that $n_i = 1\ for\ (N-1)$ nodes and for one node it is equal to $(N-1)$. It is possible when all selected links are connected to one node only.

When $1 \leq t_i \leq 2$ for $\forall i,\ i = 1,\ 2,\ \ldots,\ N$, the minimum spanning tree will have some modifications and changes. Further the number of arcs in the spanning tree with index restriction will still be $(N-1)$, the total index of all selected arcs will be still $2\,(N-1)$. Therefore only two nodes will have an index 1 and the remaining will have the index 2. Thus the minimum spanning tree under the index restriction will be like a path between two nodes i and j passing through all the remaining nodes of the given network.

It also implies that all nodes with $t_i > 2$ will require modification, i.e., some arcs will be removed, and they will be replaced by some new arcs, which can reduce the imbalance in index restriction at various nodes. The total weight of the tree will increase.

Basic and non-basic arcs: A link that has been selected as a member of the MST is said to be a basic link. Mathematically, it is a link for which $x_{ij} = 1$, i.e., it belongs to the MST. If $x_{ij} \neq 1$, then the link (ij) is non-basic, and it is not a part of the MST. Note that only for (N –1) links, $x_{ij} = 1$.

High and low index nodes of the MST: The index value of a node is a count of the number of links emanating from that node. If the number of basic links emanating from a node is greater than 2, that node becomes a high index node and if the number of basic links is 1, it becomes a low index node, except for the two nodes from where the path begins and ends.

Neighbouring node: A node i is said to be a neighbouring to node j, if the two nodes i and j are connected by a single link, i.e., $x_{ij} = 1$.

Neighbouring links: These are links that emanate from a neighbouring node. Note that in a completely connected network, all nodes are neighbouring nodes as all pairs of links are directly connected by a single link.

3.3.3 Index balancing process by network reconstruction

Theorem 3.1 Adding the same constant to all links emanating from the same node does not change the relative merit of any selected link in the MST.

Proof Consider an '*n*' node completely connected network. In such a network, from each node (*n* – 1) links will be emanating from it. In a connected network, this number can be less than (*n* – 1). Consider these (*n* – 1) links as shown in Figure 3.1.

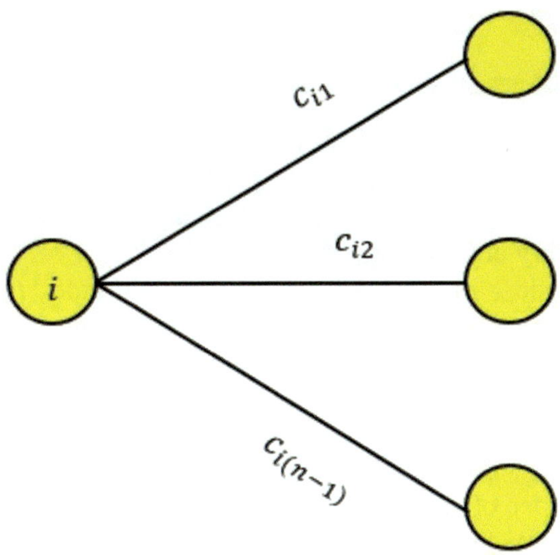

Figure 3.1 Links emanating from a typical node

Adding a constant '*d*' to all links will generate a network shown in Figure 3.2.

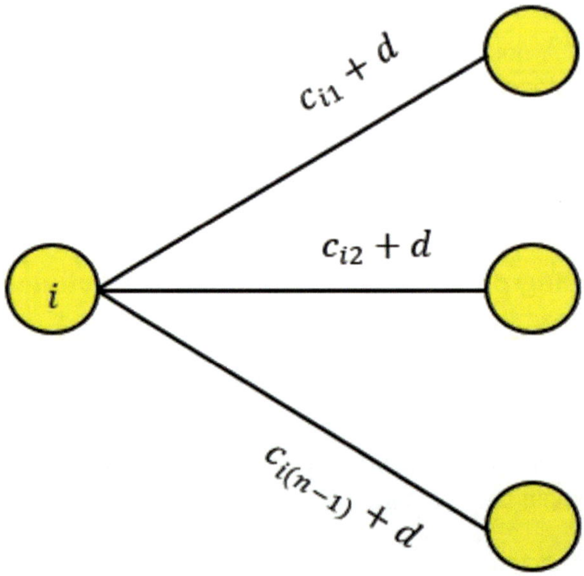

Figure 3.2 Reconstructed link weights.

Note that when link weights are altered for all links emanating from a node, the link weights will have consequences on both sides of the arcs, i.e., link weight distribution of two nodes will be affected. Therefore, two separate considerations are necessary at both ends of a link.

Consider a typical node *k* in the MST, where the node index $t_k > 2$, therefore the node *k* is causing the imbalance. However, using the greedy approach the link (kp) was selected, which is currently a basic link. Suppose, when a constant is added to all links at node *k*, relative merits of all links emanating from the node *k* are unaltered. However, when a constant is added to the link (kp), it also has a consequence at the node '*p*'. It means the link (kp), which was a member of the MST was included in the MST as the weight with the link (kp) was less than the link (pl), which made the node $'p'$ a low index node, and node *k* as the high index node.

If the constant '*d*' to be added to all links emanating from the node 'k' is such that it can make the weight equal to the link weight associated with the link (pl), then we have an alternative for selection. It can either be (kp) or (pl). Now if the link (kp) is replaced by the link (pl), the node index at node k will reduce by one and the index at node $'p'$ will go up by one, improving the index balance. This step can be repeated and all nodes with high index and low index will attain a balance.

Proof for optimality:

Let the length of the minimum MST be denoted by L(MST) and let the length of any other MST, excluding the minimum be denoted by L'(MST). It is a sum of (n – 1) selected link weights forming a spanning tree in the given 'n' node network. By using the index balancing theorem for k number of times, let $d_i,\ i = 1,\ 2,\ \ldots,\ k,$ be the weights added for each modification.

Therefore, at the end of first application of the index balancing theorem, we have:

$$L(\mathrm{MST}) + d_1 \leq \min\{L'(\mathrm{MST}) + d_1\} \tag{3.3}$$

Note that eqn (3.3) holds as eqn (3.4) holds by the definition,

$$L(\mathrm{MST}) \leq\ L'(\mathrm{MST}) \tag{3.4}$$

Similarly, at the end of the second application,

$$L(\mathrm{MST}) + d_1 + d_2 \leq\ \min\{L'(\mathrm{MST}) + d_1 + d_2\}$$

and after the kth modification, we have:

$$L(\mathrm{MST}) + d_1 + d_2 + \cdots + d_k \leq \min\{L'(\mathrm{MST}) + d_1 + d_2 + \ldots + d_k\}. \tag{3.5}$$

The additional weights just create alternatives and do not change relative merits of a given MST.

Consideration at the other end:

Note when a constant has been added to the link (k, p) to create an alternative, the length of the link (k, p) is changed and has consequence on relative merits at the node 'p'. However, all these affected links have no place in the MST, as they are non-basic. We have just created an alternative, and for maintaining the interest of index balance, we select the alternative link (p, l).

Therefore, weights can be modified, and index balance can be achieved.

3.3.4 Index imbalance for the MST of a given network to become an index restricted MST is always even

Since, the MST of an 'N' node connected network will be comprised of (N – 1) links and these links will have a total node index value 2(N – 1), which is an even number for all *N*. Hence, every application of the index balancing theorem will improve index balancing by 2, i.e., lowering the index at one node and making the index go up by 1 at some other node, which was a low index node.

Theorem 3.2

For any MST solution of a given network, the number of basic links emanating from a node '*k*' can be altered by adding a constant to all links emanating from that node.

Proof

Consider a node *k*, which is a high index node. It has more than one neighbouring nodes and let two neighbouring nodes be i and j as shown in Figure 3.3. It has three basic links and let two of these basic links be denoted by (k, i) and (k, j). The link (i, j) is not basic.

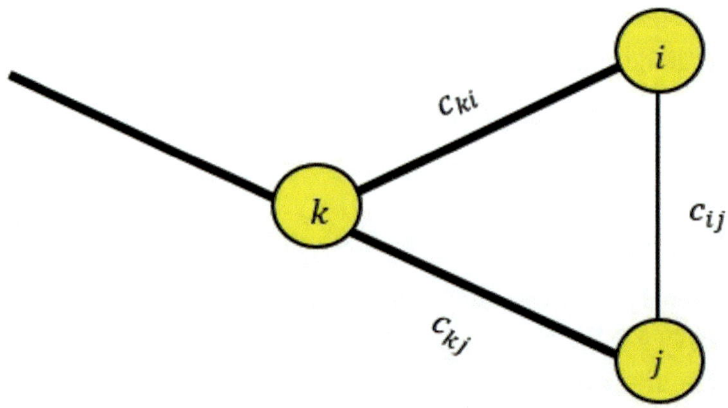

Figure 3.3 Two neighbouring nodes i and j to node *k*.

If node index for node '*k*' is high, we add an appropriate weight '*d*' to the basic links emanating from the node *k*, such that the modified weight becomes equal to the weight of the link (i, j). It is shown in Figure 3.4.

This modification can alter the basic arcs as shown in Figures 3.5 or 3.6 below. In Figure 3.5, an alternative has been created for the link (i, j).

Similarly, alternative can be created for the link (k, i) is possible, depending on the link weights.

These illustrations are intended to show as how an index of a high index node can be lowered and how to increase the index of a low index node.

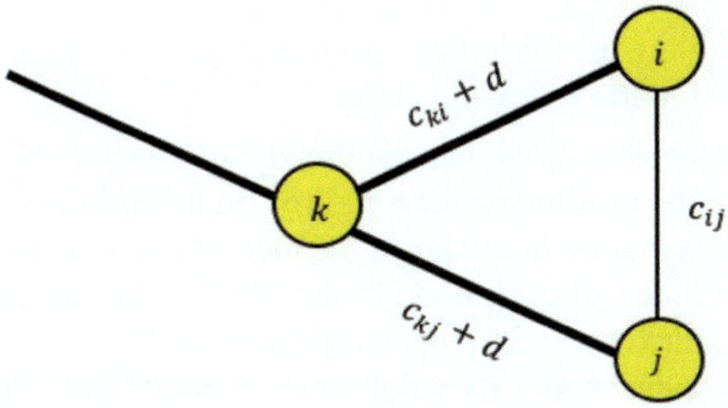

Figure 3.4 Modified weights by adding a constant '*d*'.

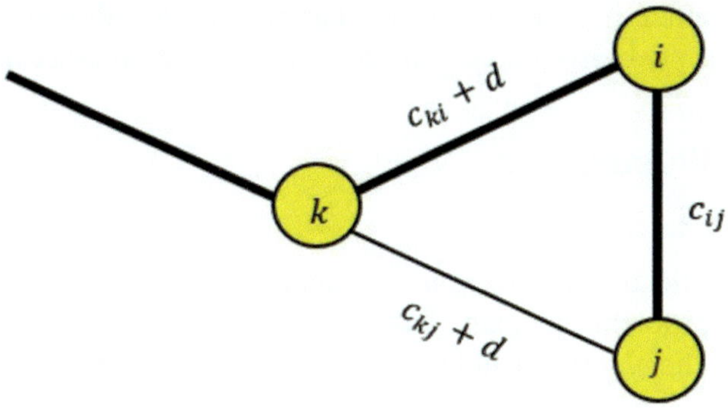

Figure 3.5 Alternative created for the link (k, j) .

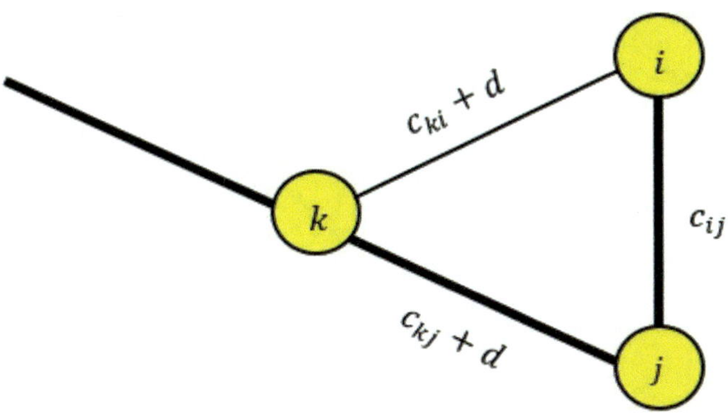

Figure 3.6 Alternative created for the link (k, i) .

3.4 The Algorithm

The algorithm for obtaining the minimum spanning tree under the node index restriction can be stated in the following steps.

Step 1: Following the steps described as the greedy approach in Section 3.2, one can obtain the minimum spanning tree. In this process, if there is a tie, we resolve that tie in favour of balance in the node index.

Step 2: Find the node index of each node in the MST. If all nodes have the desired node index, go to Step 4, else go to Step 3.

Step 3: Select a high index node, its neighbouring nodes and identify the quantity '*d*' to reduce the index at the high index node by one and increase the index of a low index node by one and resolve the tie in favour of balancing the node index. Go to Step 2.

Step 4: The MST under the index restriction has been obtained. Identify this path between two given nodes and passing through all other nodes.

3.5 Numerical Illustrations

3.5.1 Illustration 1

Consider a simple illustration in Figure 3.7 and link lengths are as given in Table 3.1. This problem has many solutions; however, the purpose is to illustrate the use of the algorithm and the index balancing theorem.

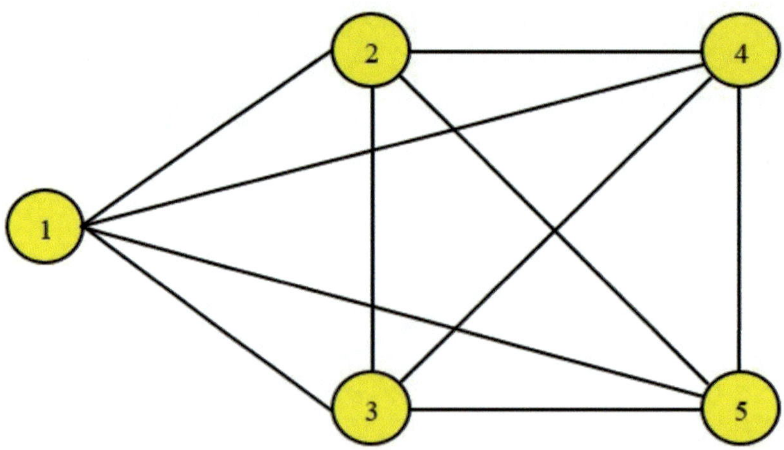

Figure 3.7 Five-node trivial network.

Link lengths for the above network are given in Table 3.1.

Table 3.1 Link weights of various links.

Nodes (*j/i*)	1	2	3	4	5
1	-	1	1	1	1
2	1	-	2	2	2
3	1	2	-	3	3
4	1	2	3	-	4
5	1	2	3	4	-

The minimum spanning tree will be as shown in Figure 3.8, i.e., all links joining the node 1 to all other nodes.

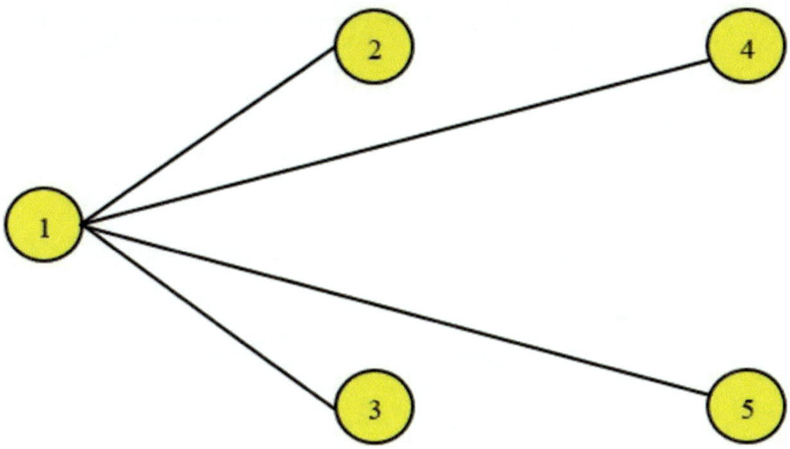

Figure 3.8 The MST of the network in Figure 3.7.

The MST in Figure 3.8 will not satisfy the requirements of the index restricted minimum spanning tree. Clearly the node 1 is a high index node and all other nodes are low index nodes. An application of the index balancing theorem is going to be useful. We select a neighbouring node to the node 1 and in this trivial network all other nodes are neighbouring nodes to the node 1. Let us arbitrarily select the node 2 as the neighbouring node. To create an alternative, we can add and modify the link weight of all links starting from node 1. This value will be 1, when 1 is added to the link weights to all links starting from the node 1, these modified weights will be as shown in Table 3.2. Note that the link (1,4) can be replaced by the link (2,4), therefore the index value of node 1 will go down by 1 and the index value of node 2 will go up by 1. Node 2 is no longer an imbalanced node. The modified MST is shown in Figure 3.9.

Table 3.2 Modified link weights.

Nodes (*j*/*i*)	**1**	**2**	**3**	**4**	**5**
1	-	1 + 1 = 2	1 + 1 = 2	1 + 1 = 2	1 + 1 = 2
2	1 + 1 = 2	-	2	2	2
3	1 + 1 = 2	2	-	3	3
4	1 + 1 = 2	2	3	-	4
5	1 + 1 = 2	2	3	4	-

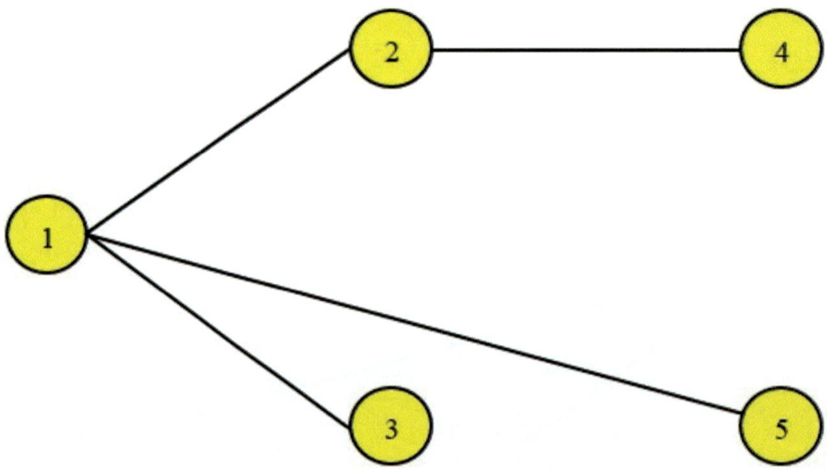

Figure 3.9 Updated index restricted modified tree.

One more application of the index balancing theorem will remove the imbalance at node 1, which is still a high index node and nodes 3 and 5 are low index nodes. The final index restricted minimum spanning tree will be as shown in Figure 3.10.

It is a path between nodes 4 and 5 passing through all other nodes. There are many other solutions to this trivial problem.

3.5.2 Numerical illustration 2

Consider a six-node connected network as given in Figure 3.11 and its link weights are given in Table 3.3.

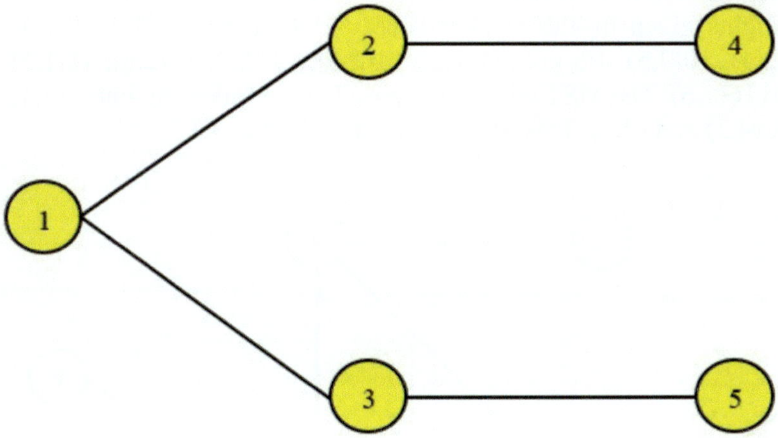

Figure 3.10 Minimum spanning tree under node index restriction.

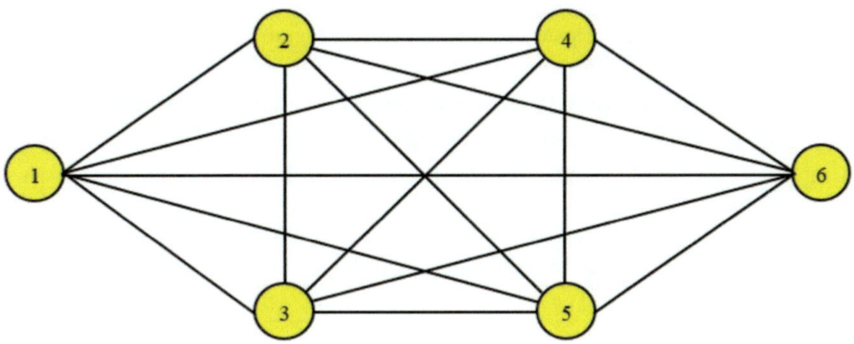

Figure 3.11 Six-node completely connected network.

Table 3.3 Link weights for the network in Figure 3.11.

From/to	**1**	**2**	**3**	**4**	**5**	**6**
1	-	11	9	9	15	16
2	11	-	14	10	10	15
3	9	14	-	6	13	11
4	9	10	6	-	9	10
5	15	10	13	9	-	8
6	16	15	11	10	8	-

Link weights in non-decreasing order for the network in Figure 3.11 from the link weights given in Table 3.3 will be: 6(3,4), 8(5,6), 9(1,3), 9(1,4), 9(4,5), 10(2,4), 10(2,5), 10(4,6), 11(1,2), 11(3,6), 13(3,5), 14(2,3), 15(1,5), 15(2,5), and 16(1,6). The MST will be composed of the following links: (3,4), (5,6), (1,3), (4,5) and (2,4). This MST is shown in Figure 3.12.

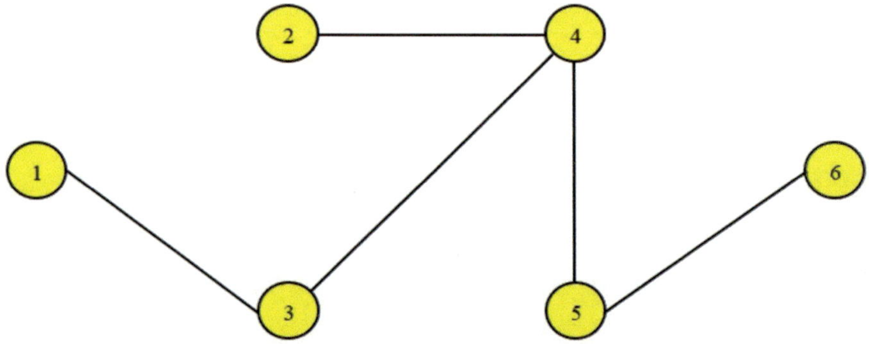

Figure 3.12 The MST of the network in Figure 3.11.

From the MST in Figure 3.12, it is clear that node 4 is a high index node and node 2 is a low index node. If we add 1 to link weights of all links emanating from the node 4, we get the modified link weights as shown in Table 3.4.

Table 3.4 Modified link weights.

From/to	1	2	3	4	5	6
1	-	11	9	9 + 1	15	16
2	11	-	14	10 + 1	10	15
3	9	14	-	6 + 1	13	11
4	9 + 1	10 + 1	6 + 1	-	9 + 1	10 + 1
5	15	10	13	9 + 1	-	8
6	16	15	11	10 + 1	8	-

From these modified weights, we can join the node 2 to form a part of the MST in three ways, i.e., remove (4,2) and replace it by (1,2), which results in an index restricted MST, as given in Figure 3.13. Here the path starts at node 2, ends at node 6 and all other nodes 1, 3, 4, and 5 have node index 2.

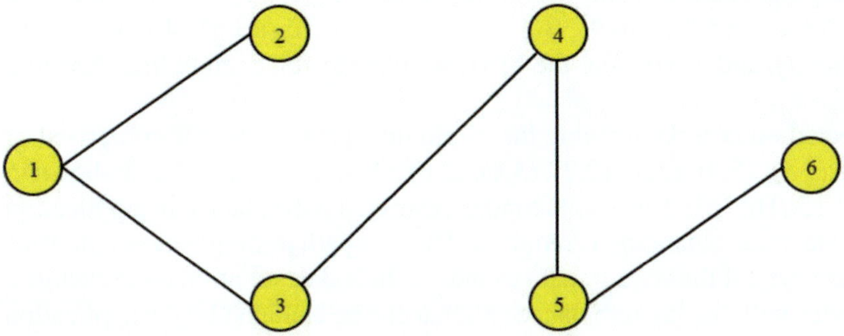

Figure 3.13 Node index restricted MST.

3.5.3 Illustration 3

Consider a nine-node 16-link connected network (Figure 3.14), where the link weights are given in Table 3.5.

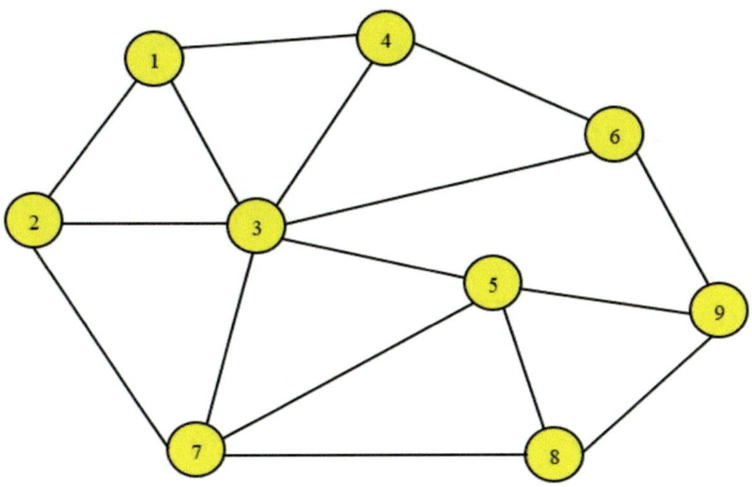

Figure 3.14 Nine-node, 16-link connected network.

Table 3.5 Link weights for the connected network.

Link	12	13	14	23	27	34	35	36	37	4,6	57	58	59	69	78	89
Wts	3	4	1	2	2	2	4	7	6	4	6	2	1	2	3	4

Arranging weights in non-decreasing order, we have 1(1,4), 1(5,9), 2(2,3), 2(2,7), 2(3,4), 2(5,8), 2(6,9), 3(1,2), 3(7,8), 4(1,3), 4(3,5), 4(4,6), 4(8,9), 6(3,7), 6(5,7), and 7(3,6). For the MST, we need to select eight links forming a tree.

Selected eight links forming the minimum spanning tree is comprised of the links (1,4), (5,9), (2,3), (2,7), (3,4), (5,8), (6,9), and (7,8). This is shown in Figure 3.15. This MST is also the index restricted MST; hence no application of the node index balancing is required. This is a path joining between nodes 1 and 6; however, if the requirement of the origin and destination node changes, some nodes will display imbalance and that can be changed by the application of the index balancing theorem.

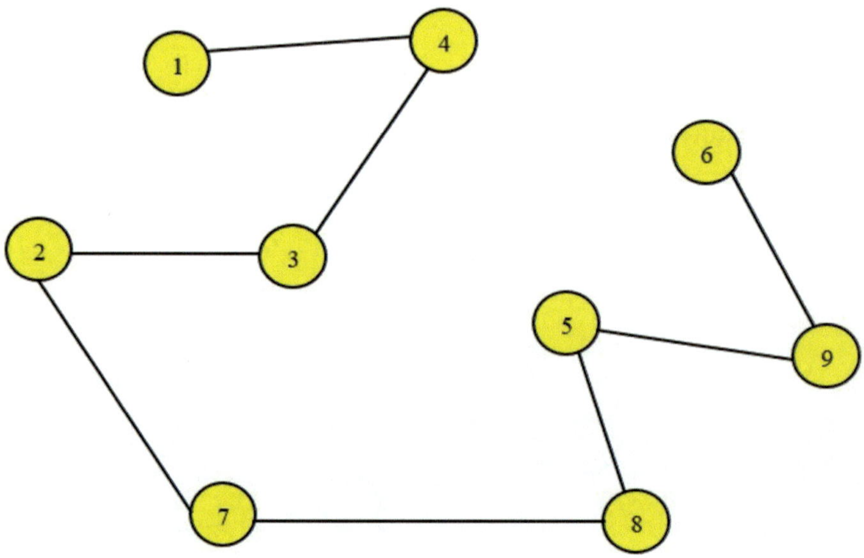

Figure 3.15 MST and node index restricted MST.

3.6 Concluding Remarks

In this chapter, a variant of a standard minimum spanning tree has been investigated, which is possible by using the node index balancing theorem. The minimum spanning tree is a classical problem in graph theory, which has applications in many real-life situations. This variant was developed by Munapo et al. (2016). This variant has applications, some of which are discussed in Chapters 4 and 5, dealing with NP Hard combinatorial problems, see Kumar and Munapo (2024).

References

G. Anupam, Deterministic MSTs, Advanced Algorithms, CMU, Spring 15-859E, pp 1-6, 2015.

R.C. Garg and S. Kumar, Shortest Connected Graph through Dynamic Programming technique, Maths. Magazine, 41 (4), pp 170-73, 1968.

R.L. Graham and H. Pavol, On the history of the minimum spanning tree problems, Annals of the History of Computing, 7(1), pp43-57, 1985.

G.L. Nemahuaser and L.A. Wolsey, Integer and combinatorial Optimization, North Holland, 1989.

M. Padberg and G. Rinaldi, A branch and cut algorithm for the solution of the large-scale symmetric travelling salesman problem, SIAM Review 33 (1), pp 60 -100, 1991.

I. Akgun and B.C. Tansel, Min-degree constrained Minimum Spanning Tree problem: New Formulation via Miller-Tucker-Zemlin constraints, Computers and Operations Research, 37(1), pp72-82, 2010.

R. Matai, S.P. Singh, and M.L. Mittal, Travelling salesman problem: An Overview of applications, formulations, and solution approaches, Chapter 1 in Travelling salesman problem: Theory and Applications by Davendra Donal (Ed.) Intech Open Access Publisher, ISBN 978-953-307-426-9, pp 1-24, 2010.

F. Peppino, De R. Floriano, S. Cesare, and F.S. Amilcare, Routing optimization in Vehicular networks: A New Approach Based on Multi-objective Metrices and Minimum Spanning Tree, International Journal of Distributed Sensor Networks, 2013. http://dx.doi.org/10.1155/2013/598675.

W. Zhaocai, H. Dongmei, M. Huajun, and T. Chengpei, A new fast algorithm for solving the Spanning Tree problem based on DNA molecules computations, Bio Systems, 114 (2013) pp1-7, 2013.

M. Li, S. Yang, J. Zheng, and X. Lu, A Euclidean Minimum Spanning Tree Based on Evolutionary Algorithm for Multi-objective Optimization, Evolutionary Computation XX(X) pp1-42, 2013.

M. Resovsky, M. Grof, D. Horvath, and V. Gazda, Revealing Ownership Structure of funds Using Minimum Spanning Tree, ACTA Physica Polonica A, 126(6) pp 1322-1326, 2014.

L.W. Beineke, R.J. Wilson, *Topics in topological graph theory, Encyclopaedia of Mathematics and its Applications, vol. 128, Cambridge University*

Press, Cambridge, p. 36, 2009. doi:10.1017/CBO9781139087223, ISBN 978-0-521-80230-7, MR 2581536

E. Munapo, S. Kumar, M. Lesaoana, and P. Nyamugure, A minimum spanning tree with node index <= 2, The ASOR Bulletin, Vol 34, No 1, pp 1-14, 2016. (www.asor.org.au)

S. Kumar, E. Munapo, M. Lesaoana, and P. Nyamugure, Is the travelling salesman problem actually NP hard?, Chapter 3 in Engineering and Technology: Recent Innovations and Research, Editor A Matani, International Research Publication House, 2016, ISBN 978-93-86138-06-4, pp 37-58.

S. Kumar, E. Munapo, M. Lesaoana, and P. Nyamugure, A minimum spanning tree-based heuristic for the travelling salesman tour, Opsearch, Vol 55, No 1, pp150-164, 2018. {DOI10.1007/s12597-017-0318-5pp1-15,2017}

S. Kumar, E. Munapo, C. Sigauke, M. Al-Rabeeah, M, The minimum spanning tree with node index <=2 is equivalent to the minimum travelling salesman tour, Chapter 8, Mathematics in Engineering Sciences: Novel Theories, Technologies and Applications (Editor) Mange Ram, CRC Press, 2019(http://www.crcpress.com) pp 227-243, ISBN 9781351266321.

J.E. Mitchell, Branch and cut algorithms for integer programming: In Christodous, Floudas and Pardalos (Editors) Encyclopaedia of Optimization, Kluwer Academic Publishers,2001, http://homepages.rpi.edu/-mitch j/bioJEM.pff

D. Nadef, Polyhedron theory and branch and bound cut algorithms for the symmetric travelling salesman problem. In Guiten and Punnen (Editors) The travelling salesman and its variations. Kluwer Dorcdrecht, pp29-116, 2002.

L. Cowen, The travelling salesman problem, Lecture 4, Comp 260, Advanced Algorithms, Tuffs University, 2011 http://www.cs.tuffs.edu/comp/260/old/ lecture4.pdf

E. Munapo, and S. Kumar, Linear Integer Programming: Theory, Applications, Recent Developments, De Gruyter Series on the applications of Mathematics in Engineering and information sciences, De Gruyter, 2022, ISBN 978-3-11-070292-7

W.L. Winston, Operations Research: Applications and algorithms, Duxbury Press 4th Edition, 2004.

S. Kumar, and E. Munapo, Innovative ways of developing and using specific purpose Alternatives for solving hard combinatorial network routing and ordered optimization problems, AppliedMath 2024, 4, 791-805, http://doi. org/10.3390/appliedmath4020042

4

Path through '*k*' Specified Nodes or *k*' Specified Links in Networks: A Reconstruction Approach

Abstract

In this chapter, a variant of the shortest route problem has been considered, where the path between the origin and the destination is constrained to pass through a given set of specified nodes or pass through given set of specified links. These variants are important as they have real-life applications, however, these constraints make the problem computationally difficult. A heuristic based on the spanning tree under the index restriction can resolve the situation to provide a quick solution, which is a near-optimal solution. Difficulties associated with this approach are independent of the number of specified elements in the set and it does not run into computational explosion.

Keywords: Specified nodes, specified links, circuits, minimum spanning tree under index restriction, partially minimum spanning tree, basic links.

Notation

1. $G(n, m)$ represents a network, where n and m are two sets of nodes and links.
2. The set of nodes $\{n\}$ has *N* number of elements representing nodes, $i = 1, 2, \ldots, N$. The set $\{m\}$ has *M* number of elements representing links (i, j) by joining the nodes i and j, $i,\ j = 1, 2, \ldots, N$ and $i \neq j$.
3. A specified node (i) is represented by $(i\ s)$, similarly a specified link (i, j) is represented by $(i\ s\ j)$.

$$x_{ij} = \begin{cases} 1, \text{ if the link } (i,j) \text{ belongs to the spanning tree} \\ 0, \text{ if the link is not a member of the spanning tree} \end{cases}.$$

4.1 Introduction

Many industrial, management and business-related situations can be represented by a network $G(n, m)$, where n and m represent the set of nodes and links joining the pair of nodes. Many of these network problems reduce to find a route joining two nodes, usually designated as the origin node and the destination node. This routing problem is a well-known classical problem in network theory. Depending on the situation, they can be transformed in the form of finding the shortest or the longest route between the starting point (origin node) and the terminal point, the destination node. Many approaches have been suggested and research interest remains active for this problem. For example, see Dantzig (1957), Bellman (1956, 1962), Peart et al. (1960), Pollak and Weibenson (1960), Beckwith (1961), Ahuja et al. (1990), Rais et al. (2014).

We discussed the shortest route problem in Chapter 2, where the reconstruction approach was introduced. In Chapter 3, we obtained the minimum spanning tree under the index restriction. Here, in this chapter, we consider a variant of the shortest route problem, which was suggested by Kalaba (1960). He pointed out a situation where a need arises for a path to pass through a set of specified nodes before reaching the destination node. When the set of specified nodes is a null set, the problem reduces to an ordinary routing problem, i.e., determination of a path between two known nodes, see Zhan (1997) and Kumar et al. (2013). When the set of specified nodes contains all nodes and the path is restricted to return back to the starting node, it becomes a typical travelling salesman problem, see Bellman (1962), Munapo and Kumar (2022), Kumar et al. (2016, 2017, 2020). More details on the travelling salesman tour will be discussed in Chapter 5 in this book.

In this chapter, we discuss two variants of the shortest route problem, when the path must pass through a set of specified nodes or through a set of specified links before arriving to the destination node. Specifically, this means that:

(1) when the set of specified nodes is not a null set, i.e., the path from the origin must pass through the specified nodes before arriving at the destination node, see Kumar et al. (2014);
(2) when the set of specified links is not a null set, i.e., the path from the origin node to the destination node must pass through the links in the specified set before reaching the terminal point; see Kumar et al. (2023).

This chapter is organized in seven sections. In Section 4.2, we discuss some of the industrial situations, where the above variants of the routing problem have applications. Section 4.3 illustrates a typical set of specified nodes and a set of specified links. Section 4.4 deals with the complexity that arises due to the constraint of specified nodes or links. In Section 4.5, we discuss a heuristic to find the path from the origin to the destination node, which is a constraint to pass through the set of specified nodes before arriving at the destination node. Section 4.6 presents the problem of finding a path through a set of specified links in a network and finally, the chapter is concluded in Section 4.7.

4.2 Industrial Applications for Specified Set of Nodes or Links

We know that nodes in a graph indicate some physical entity, like a city in a country and links joining any pair of nodes are a kind of reflection of the relationship between those entities. For example, the points of interest for a tourist may be represented by nodes, and the associated link joining them may represent time or cost of travel. The link weight is used to quantify the relationship between those nodes. Interpretation associated with an entity and relationship changes from one situation to another. Here we attempt to identify some industrial situations, where specified nodes or links may have meaningful interpretations. A few illustrative cases are presented below.

4.2.1 Guided tour companies

Guided tours are a multi-billion-dollar business all over the world. Tours are planned and programmed meticulously. However, the interest of the tour company remains to be able to extract a little more from the tourist and in return give them more. It is for this reason; they often develop some more possibilities for the arriving tourist to enjoy a few more pre- or post-tours of short duration. The idea is to cover a bit more at a marginal cost. This is just like the concept of specified nodes or links to be covered before returning to the point of origin. This is an attractive offer for the tourist as well as for the tour company. The idea from the tourist's point of view is to do a bit more before the return and the tourist company wants to make their stay more memorable by giving them more sightseeing.

4.2.2 National planning

In the real world, our activities are surrounded by protean systems, i.e., systems with a characteristic that keep them changing with time, see Kumar

(1995), Kumar et al. (1999a,b), Kumar (2005, 2006). National planning and improvements for the future are common situations often faced by various management systems. For example, traffic congestion is one illustration, which often requires upgrades. Supply of funds is limited for any organization; therefore, management often faces the problem of prioritizing and selecting a few links to work on them in the limited funds, and to avoid excessive disruption of service to accommodate construction work. The problem is where and what routes can be improved. The selection of routes and a section on that route of a specific bottleneck forms a set of specified links selected for improvements.

4.2.3 Delivery points

It is well known that manufactured goods travel various distances from the factory where they were produced to their destination, the consumer. As an example, consider a department store with many outlets in a large cosmopolitan city. Each store cells many bulky goods such as fridges, washing machines, dishwashers, etc. which usually are not collected by the buyers. These customers often request the delivery and installation of these bulky goods at their residences. Usually, each store will not hold separate inventory at the store but sold items from different outlets are delivered from a central warehouse to various homes, where a sale has been identified. Each household residence is situated on a street in some residential area, and these streets, where deliveries for a particular day are scheduled, form the set of specified links. The warehouse is the origin node, and the delivery truck depot is the destination. The specified links are the delivery links (roads indicating the residential address on a particular day). This set of specified links changes each day as customer set changes each day.

Similar situations can be identified in many other distribution centres, for example, farm goods are transported to different warehouses, manufactured goods are transported from the production plant to various warehouses, etc. The problem of passing through a set of specified links can be identified in various ways in different industries involved in production, distribution and after-sales service.

4.3 Typical Illustrations of Specified Nodes or Link in $G(n, m)$

Specified nodes are given, they form a subset of the given node-set $\{n\}$. The path from the origin node must pass through these specified nodes before it arrives at the destination node *N*. The specified nodes are illustrated in Figure 4.1 (a), where specifies nodes are shown with the letter '*s*' along with the node number. Nodes 1 and *N* are the origin and destination nodes.

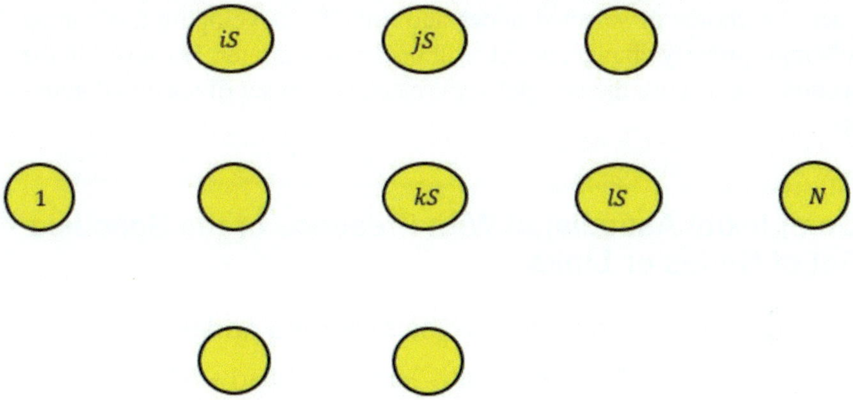

Figure 4.1 (a) Specified nodes are $(i,\ j,\ k,\ l)$.

Similarly, the set of specified links is shown in Figure 4.1 (b).

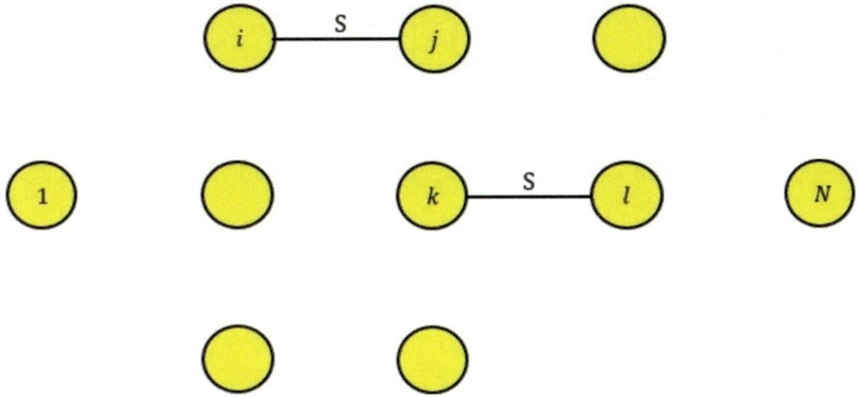

Figure 4.1 (b) Specified links $\{(ij)\,,\ (kl)\}$.

In Figure 4.1 (a), four nodes $(i, j,\ k, l)$ are forming the set of specified nodes and in Figure 4.1 (b), links (i, j) and (k, l), form the set of specified links. In general, we may assume that the number of specified nodes is given by *r*, where $0 < r < n$ (the value of *r* in Figure 4.1 (a) is 4). Specified links are denoted by '*p*', where $0 < p < L$. The number of specified links in Figure 4.1 (b) is 2, denoted by (i, j) and (k, l). The number of specified nodes and specified links is governed by the situation, where such path is required.

The spanning tree of the graph $G(n, m)$ will have a total number of (N – 1) links and N number of nodes. When specified nodes or links set is not a null set, the choice of selected nodes and links for developing a spanning tree will appropriately change, which will be discussed later. However, in the next section, we discuss the complexity created by the set of specified nodes or links.

4.4 Complexity Associated With Presence of the Specified Set of Nodes or Links

4.4.1 Complexity associated with the specified nodes

Complexity of the travelling salesman is known; similarly, when a path is restricted to pass through the set of specified nodes, complexity increases with an increase in the number of specified nodes. This complexity can be described by considering a few specific cases, as discussed below:

1. Consider one specified node denoted by r_{1s}.
   Possible path: $O \to r_{1s} \to N$
   No. of shortest route problems solved $= 2$
   No of new problems solved $= 2$
   No of evaluations $= 1$
   Complexity pattern 2(Number of specified nodes *r*) + $r_{C_2} = 2(1) + 0 = 2$
2. Consider two specified nodes r_{1s} and r_{2s}
   Possible paths: $O \to r_{1s} \to r_{2s}$ -> *N* or $O \to r_{2s} \to r_{1s} \to N$
   No of problems solved $= 3 + 2 = 5$
   No of evaluations 2
   Complexity pattern $= 2(2) + 2_{C_2} = 5$
3. Consider three specified nodes, r_{1s}, r_{2s}, and r_{3s}. There are six possible paths: $O \to r_{1s} \to r_{2s} \to r_{3s} \to N$ or $O \to r_{1s} \to r_{3s} \to r_{2s} \to N$ or $O \to r_{2s} \to r_{1s} \to r_{3s} \to N$ or $O \to r_{2s} \to r_{3s} \to r_{1s} \to N$ or $O \to r_{3s} \to r_{1s} \to r_{2s} \to N$ or $O \to r_{3s} \to r_{2s} \to r_{1s} \to N$
   No of problems solved $= 9$
   No of evaluations $= 6$
   Complexity pattern $= 2(3) + 3_{C_2} = 9$

In general, when 'k' nodes are specified, the complexity will be given by $= 2(k) + k_{C_2}$.

4.4.2 Complexity associated with 'k' specified links

Consider the set of specified links contain 'k' number of links, where $k > 1$. If these specified links do not form a circuit, we need $(N - k - 1)$ additional links to form a connected graph. Therefore, after selecting $(N - k - 1)$ additional links, one obtains a connected spanning tree of the network $G(n,m)$. These additional $(N - k - 1)$ links can be selected by following Kruskal (1956) steps, which can be slightly modified for our purpose. Since the final tree will have 'k' specified links and $(N - k - 1)$ selected links, it will not be a minimum spanning tree, we call this tree a 'partially minimum spanning tree' (PMST). In some cases, depending on the configuration of the specified links, this graph may have even circuits. After slight modification, the Kruskal (1956) steps for the situation presented in this chapter will be:

Step 0: Include the specified links in the set of selected links.

Step 1: Arrange the remaining $(M - k)$ number of non-specified links in non-decreasing order of their weight.

Step 2: Pick the smallest link. Check if it forms a cycle with the selected links, which initially are the set of specified links. Remember, the specified links may have a cycle among them. In subsequent selections, one must check with the selected non-specified links and the given specified links, that the new link does not form a cycle, include this link, else discard it.

Step 3: Repeat Step 2, with the next link until total number of links i.e., specified 'k' and non-specified $(N - k - 1)$ totalling equal to $(N - 1)$. If the specified links form a cycle, for each cycle, one additional link will be required to form a partially minimum spanning graph, which may have some loops also.

4.5 Mathematical Model of the Shortest Path Through K Specified Nodes

The path through 'k' specified nodes was considered by Saksena and Kumar (1966). Computational complexity in their approach was a function of the number of elements in the specified node set. Here we discuss an approach that uses the spanning tree under the node index restriction, which avoids explosion in the computational complexity.

4.5.1 Algorithmic steps to find the path through the nodes that belong the set of specified nodes

Let S denote the set of specified nodes and let the number of elements in this set be denoted by 'r', where 'r' is an integer such that $0 \leq r \leq (N-2)$, N is number of nodes in the given network $G(n, m)$.

Step 1: Find the shortest path between the origin and the destination node, without imposing any condition of passing through the nodes that belong to the set of specified nodes. This path can be obtained by any known method. Let the nodes on this path be denoted by the set SP (shortest path). If the set S is a subset of the set SP, terminate the search and go to Step 5.

Step 2: Find the MST of the given network by any known greedy approach. This minimum spanning tree will have all nodes specified as well as those which do not belong to the specified set 'S'. Trace a path joining the origin node to the destination node, if it passes through all nodes in the set of specified nodes, once again terminate the search and just remove branches leading to nodes that are not members of the set of specified nodes.

Step 3: If a specified node has an index number 1, do one of the following:

(1) See, if the index value of this specified node can be increased by 1, by using the index balancing theorem discussed in Chapter 3.

(2) Link weight of all links from this specified node, where the index value is 1 will be more, as they have been excluded from the MST. However, since loops are possible, this specified node can be included by visiting this specified node and returning to the same node to join the rest of the path. This will be a loop.

Step 4: Delete all branches leading to nodes that are not in the set of specified nodes.

Step 5: Identify the path through the set of specified nodes.

4.5.2 Numerical illustration

4.5.2.1 Example 1

Consider the network used by Saksena and Kumar (1966), shown in Figure 4.3, and link weights are given in Table 4.1. Let the set of specified nodes S be $\{3, 5, 7, 9\}$.

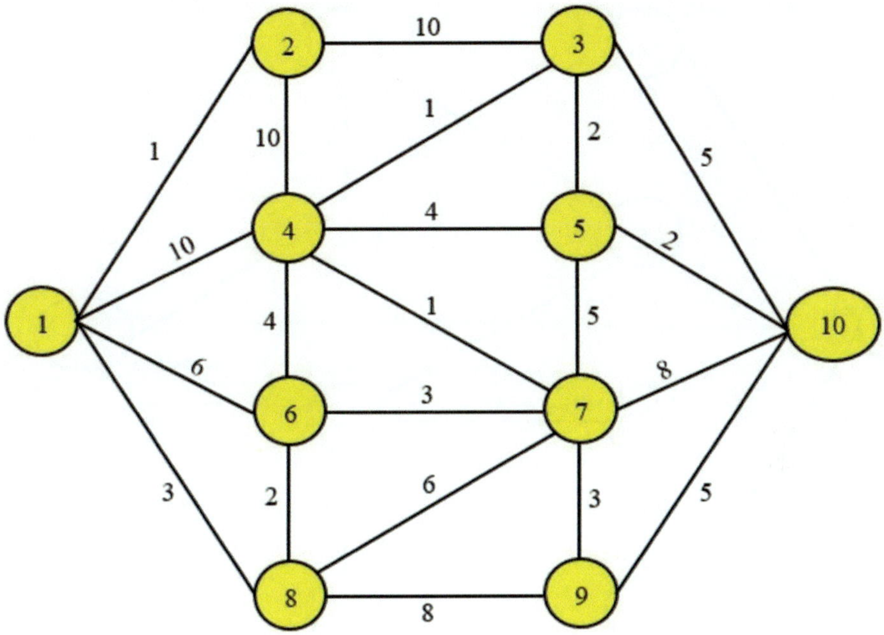

Figure 4.3 Network for the specified nodes.

Table 4.1 Link weights for the network in Figure 4.3.

Nodes	1	2	3	4	5	6	7	8	9	10
1	-	1	∞	10	∞	6	∞	3	∞	∞
2	1	-	10	10	∞	∞	∞	∞	∞	∞
3	∞	10	-	1	∞	∞	∞	∞	∞	5
4	10	10	1	-	4	4	1	∞	∞	∞
5	∞	∞	2	4	-	∞	5	∞	∞	2
6	6	∞	∞	4	∞	-	3	2	∞	∞
7	∞	∞	∞	1	5	3	-	6	3	8
8	3	∞	∞	∞	∞	2	6	-	8	∞
9	∞	∞	∞	∞	∞	∞	3	8	-	5
10	∞	∞	5	∞	2	∞	8	∞	5	-

The shortest path without any restrictions can be obtained using any method which is as follows:

$1 \rightarrow 8 \rightarrow 6 \rightarrow 7 \rightarrow 4 \rightarrow 3 \rightarrow 5 \rightarrow 10$, total cost $=14$.

Note that the nodes on the shortest path include three of the four specified nodes, i.e., 3, 5, and 7. However, the specified node 9 is missing from the shortest path route.

Next, we find the MST of the given network, which is given in Figure 4.4. It is comprised of the following 9 links. {(1,2), (1,8), (6,8), (7,6), (5,10), (3,5), (4,3), (7,4), (9,7)}.

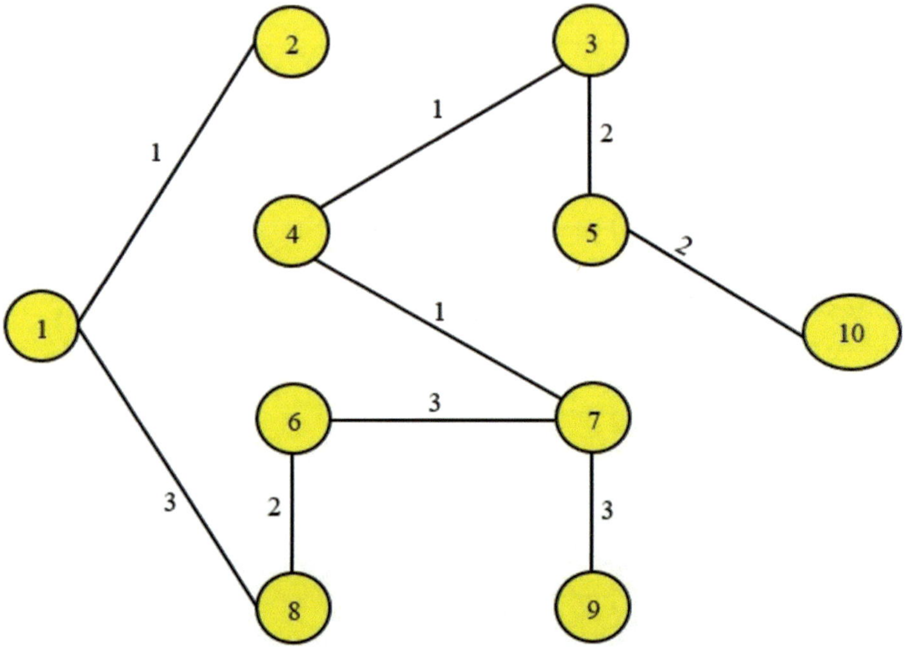

Figure 4.4 MST of the network in Figure 4.3.

Increasing index from 1 to 2 of the specified node 9 will require increasing the index of node 8 and the cost associated will be high. However, node 9 can be reached from node 7 and return to node 7 will result in the path with a loop passing through all specified node, as given below:

$1 \to 8 \to 6 \to 7 \to 9 \to 7 \to 4 \to 3 \to 5 \to 10$. Total cost for this path will be 20 units.

4.5.2.2 Example 2

Consider that we are required to find the path from node 1 to 10 passing through nodes 2, 3, 5, 7, and 9, i.e., the set of specified nodes is given by $\{2, 3, 5, 7, 9\}$.

Once again, the shortest path misses out two specified nodes, i.e., node 2 and node 9. Once again, increasing the index value of these two nodes, which is currently 1 to two, will increase the cost compared to the loop created between nodes 7 and 9 and also a loop created between nodes 1 and 2. The final path from the Figure 4.4 will be:

$1 \to 2 \to 1 \to 8 \to 6 \to 7 \to 9 \to 7 \to 4 \to 3 \to 5 \to 10$ and the total cost will be 22 units. The two loops on this path are: $1 \to 2 \to 1$ and $7 \to 9 \to 7$.

4.6 Mathematical Model of the Shortest Path Through K Specified Links

We use the following notation to develop a mathematical model for the above problem.

Let C_{ij} = the link weight associated with the link (i, j), $i = 1, 2, \ldots, n$; $j = 1, 2, \ldots, n$; $i \neq j$. Note that if there is no link from any node i to node j, that $C_{ij} = \infty$ or equivalently the corresponding variable $x_{ij} = 0$. In general, the variables will be defined as given below:

$$x_{ij} = \left\{ \begin{array}{l} 1, \text{ if the link } (ij) \text{ is used to form the path} \\ 0, \text{ if the link } (ij) \text{ does not form the path} \end{array} \right\}$$

The model can be represented as:

$$\text{Min } Z = \sum_{(ij)=1}^{L} C_{ij} x_{ij}.$$

Subject to:

$\sum_{(1,j)} x_{(1,j)} = 1$ (Indicates that one link will be used from the origin node 1).

$\sum_{(j,n)} x_{(j,n)} = 1$ (Indicates that one link will be used to reach the destination node n).

The value of

$x_{(ij)S} = 1$, for each specified links (i^s_j), where $s = 1, 2, \ldots, r$. These links must be included in the required path.

For nodes, other than the origin and the destination, and those which are free of the specified links, must satisfy the following:

$\sum_i x_{ij} = 0$ or 2, means that the shortest path passing through specified links will either not pass through a link or will enter from one node and leave that node from the other link.

For nodes from where specified links emanate or end, must satisfy:

$\sum_i x_{ij} = 2$ or 4 means that in addition to the specified link, either three or one more link will be emanating from this node as part of the path.

Note several constraints involve on the RHS having two values like 0 or 2 and 2 or 4.

These constraints will require further modification, which is explained below:

For example, consider the constraint:

$\sum_i x_{ij} = $ 2 or 4, which can be modified as

$$\sum_i x_{ij} = 2y_1 + 4y_2,\ y_1 + y_2 = 1,\ y_1, y_2 = 0 \text{ or } 1.$$

This will result in a large cumbersome model; hence some special approach is desirable, which is discussed in Section 4.6.1.

4.6.1 Further clarification of the modified selection process

The nodes forming the specified links are $(i,\ j)$ and $(k,\ l)$ in Figure 4.1(b), however, when *r* disconnected specified links are given, the number of nodes associated with the specified links will be at most equal to 2*r*. For example, in Figure 4.1(b), total node index of the nodes associated with the specified links will be equal to 4. This node index can be less than 2*r*, when these specified links are not disconnected, for example, as shown later in an illustrative example in Section 4.6.2, where the specified links form a circuit. In the final connected graph, all nodes will have node index $> = 1$. The total index value of all nodes will be $\geq 2(n-1)$. By the application of the index balancing theorem, we try to attain balance among the index of all nodes, which varies depending on the configuration of the specified links. For the illustrative example 4.6.2.1, node index will be equal to 2, except the origin and the destination node, and this index can be more than 2, i.e., 4. For developing a path joining the origin to the destination, passing through all nodes associated with the specified links, one may be forced to add more links. The node index for the nodes associated with the specified links will be 2 or 4 but not 3. The index value 4 indicates formation of a cycle.

Note that the node index balancing theorem is not applicable to the specified links, as they are never to be removed from the final graph. For the graph in Figure 4.3, the index will be equal to 2 or 4 but not three. For the node of the specified link, one link will be the given specified link, i.e., the link $(,\ j)$ at nodes i and j and the link (k, l) for the nodes k and l. The index at the origin and destination node will be 1. All other links that are not member of the path can be removed from the system.

The index balancing theorem discussed in Chapter 3, which states that, 'adding the same constant to all arcs emanating from the same node does not change the relative merit of any selected arc in the PMST (Partial Minimum Spanning Tree) obtained by the greedy approach. Since the PMST is known, the index for each node is also known, and therefore, high and low index nodes can be easily identified. The balancing index commences its balancing task by adding constants and bringing index of high index nodes low and taking index of low index node to the required level'. Since the presence of the specified links is not based on any selection, the index balancing theorem is not necessarily applicable to these links.

When required path cannot be established by the application of the node balancing theorem, additional links may have to be added but these additional links develop circuits and make node index of some nodes more than 2 to 4, an even number. All unnecessary links not on the path can be removed.

4.6.2 Numerical illustrations

In this section, we discuss two numerical illustrations with different configurations of the specified links.

4.6.2.1 Numerical illustration 3

Let us reconsider the network $N(10, 21)$ as shown in Figure 4.3 where node 1 is the origin node and 10 is the destination node. The link weights are indicated on the links as well as given in Table 4.1.

Let the set of specified links be $\{(2,3), (4,5), (6,7)\}$. Since these links are specified, they will remain part of PMST in the remaining analysis. Here onwards we will show these specified links with the letter '*S*' to indicate that they are specified links. They must form a part of the path to be identified.

4.6.2.2 Mathematical model of the above problem

For the above 10 node 21 links network, we define 21 variables. These are binary variables, associated with the links as follows:

$x_{ij} = 1$, if the link joining nodes (i, j) is part of the path, else it is zero.

Using the above, a mathematical model can be developed as given below. It will have 21 terms, one term for each link.

$$\text{Min } Z = x_{12+}10x_{14} + 6x_{16} + 3x_{18} + \cdots + 5x_{310} + 2x_{510} + 8x_{710} + 5x_{910}$$

Subject to

(1) Constraints for the origin and the destination nodes:
$x_{12+}x_{14} + x_{16} + x_{18} = 1$ (Origin node)
$x_{310} + x_{510} + x_{710} + x_{910} = 1$ (Destination node)

(2) Constraints representing nodes free of specified linksNodes 8 and 9 are the only two nodes in this category. The constraints for these two nodes

are given below for nodes 8 and 9, respectively: $x_{18}+x_{68}+x_{78}+x_{89} = 0$ or 2 and $x_{89}+x_{79}+x_{910} = 0$ or 2

These two constraints can be replaced by the following two sets of constraints, given below:

$$x_{18}+x_{68}+x_{79}+x_{89} = 0\,y_1+2y_2$$
$$y_1+y_2 = 1,\ y_1, y_2 \text{ are 0 or 1}$$

Similarly, we rewrite the constraint representing the node 9 as follows:

$$x_{89}+x_{79}+x_{910} = 0\,y_3+\ 2y_4$$
$$y_3+y_4 = 1, y_3, y_4 \text{ are 0 or 1 variables.}$$

(3) The third set of constraints will arise from the nodes that start or end at the specified links. In this category, there are 6 nodes, which are: 2, 3, 4, 5, 6, and 7. These constraints are given by:
Node 2:

$$x_{12}+x_{24}+x_{23} = 2y_5+4y_6$$
$$y_5+y_6 = 1, y_5, y_6 \text{ are 0 or 1.}$$

Node 3:

$$x_{23}+x_{43}+x_{53}+x_{310} = 2y_7+4y_8$$
$$y_7+y_8 = 1,\ y_7, y_8 \text{ are 0 or 1.}$$

Node 4:

$$x_{14}+x_{42}+x_{43}+x_{46}+x_{47}+x_{45} = 2y_9+4y_{10}$$
$$y_9+y_{10} = 1,\ y_9, y_{10} \text{ are 0 or 1.}$$

Node 5:

$$x_{45}+x_{35}+x_{57}+x_{510} = 2y_{11}+4y_{12}$$
$$y_{11}+y_{12} = 1,\ y_{11}, y_{12} \text{ are 0 or 1.}$$

Node 6:

$$x_{16}+x_{68}+x_{67}+x_{64} = 2y_{13}+4y_{14}$$
$$y_{13}+y_{14} = 1,\ y_{13}, y_{14} \text{ are 0 or 1.}$$

Node 7:

$$x_{67}+x_{47}+x_{87}+x_{79}+x_{57}+x_{710} = 2y_{15}+4y_{16}$$
$$y_{15}+y_{16} = 1,\ y_{15}, y_{16} \text{ are 0 or 1.}$$

(4) Conditions on the specified links

$$x_{23} = x_{45} = x_{67} = 1.$$

(5) Conditions on variables

All other $x_{ij} =$ 0 or 1 for all ij that correspond to a link

$$y_1, \ldots, y_{16} = 1 \text{ or } 0.$$

In the above 10 node networks; the PMST will be comprised of 9 links, as the specified links are free of a circuit. Since the specified links are 3, we need to select 6 more links to make a PMST of the network in Figure 4.3. The above network is comprised of 21 links, 3 of them are specified. The remaining 18 links are arranged in non-decreasing order as given below, where equal weights are just arranged in the order of the first node of that link.

Links in non-decreasing order (weight, link) are:$\{1(1,2)_1$,$1(3,4)_2$, $1(4,7)_3$, $2(3,5)_4$, $2(5,10)_5$, $2(6,8)_6$, $3(1,8)_7$, $3(7,9)_8$, $4(4,6)_9$, $5(3,10)_{10}$, $5(5,7)_{11}$, $5(9,10)_{12}$, $6(1,6)_{13}$, $6(7,8)_{14}$, $8(7,10)_{15}$, $8(8,9)_{16}$, $10(1,4)_{17}$, $10(2,4)_{18}\}$; and the specified links are: $\{10\ (2,3), 4(4,5), 3(6,7)\}$.

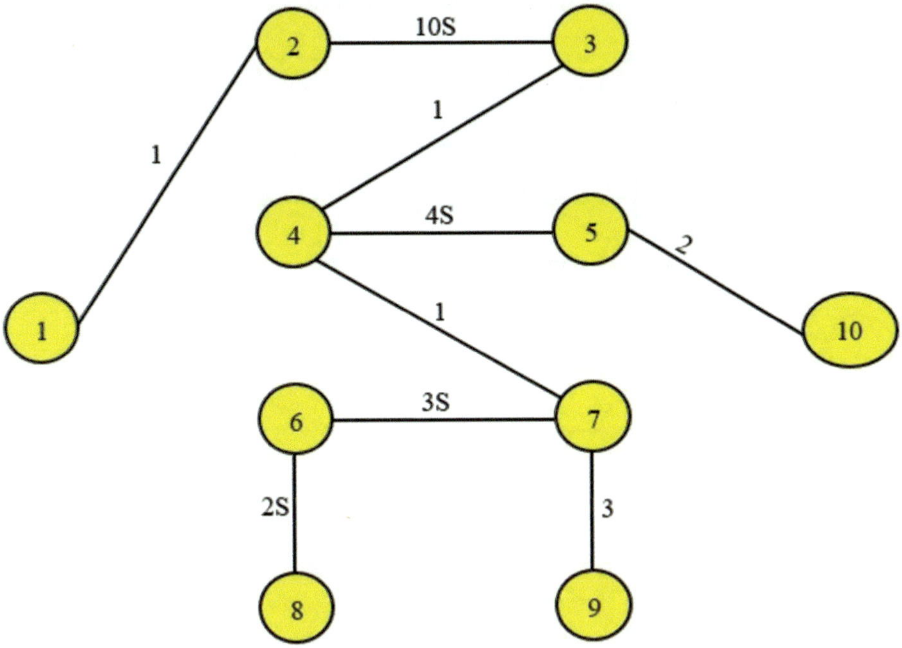

Figure 4.5 MST of the given graph.

First three links {(1,2), (3,4), (4,7)} can be selected for inclusion in the PMST but the next link (3,5) forms a loop, hence is not included. Next two links (5,10) and (6,8) are included in the PMST. Finally, the link (7,9) is included in the PMST. This PMST is shown in Figure 4.5.

Note from Figure 4.5, we are looking for a path from nodes 1 to 10 passing through the specified links (2,3), (4,5) and (6,7). Links (6,8) and (7,9) can be removed as they are not necessary for a path from 1 to 10. After removing these two links, nodes 4 and 6 have node index of 3 and 1, which is not even. Hence by adding the link (4,6), index of all nodes will be balanced, and the path becomes:

Path : $1 \rightarrow 2 \rightarrow 3 \rightarrow 4 \rightarrow 7 \rightarrow 6 \rightarrow 4 \rightarrow 5 \rightarrow 10$ with a loop $4 \rightarrow 7 \rightarrow 6 \rightarrow 4$

Total cost $= 1 + 10(S) + 1 + 1 + 3(S) + 4 + 4(S) + 2 = 26.$

The path through the specified links is comprised of a loop, which is typical in a path through specified links.

4.6.3 Numerical illustration 4

Let us reconsider the network $N(10, 21)$ in Figure 4.3 with a slightly different configuration of the specified links as shown in Figure 4.6.

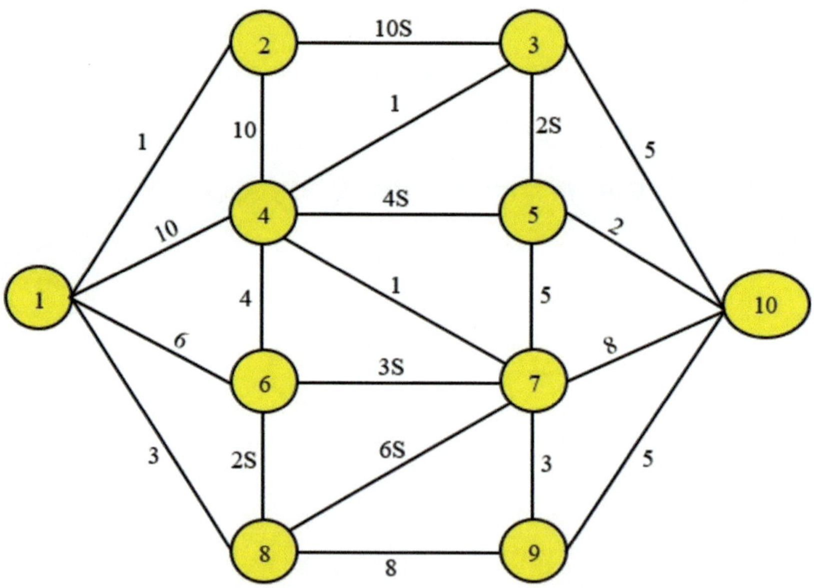

Figure 4.6 Network with specified nodes forming a disconnected set.

In this illustration, we consider 6 links forming the set of specified links. We assume they are: {(2,3), (3,5), (4,5), (6,7), (7,8), (6,8)}. To form a PMST, we need to select at least $(9-6)=3$ more links. The remaining non-specified 14 links are arranged in non-decreasing order as given below:

$\{1(1,2)_1,\ 1(3,4)_2,\ 1(4,7)_3,\ 2(5,10)_4,\ 3(1,8)_5,\ 3(7,9)_6,\ 4(4,6)_7,\ 5(3,10)_8,\ 5(9,10)_9,\ 6(1,6)_{10},\ 8(7,10)_{11},\ 8(8,9)_{12},\ 10(1,4)_{13},\ 10(2,4)_{14}\}$

Since the number of specified links is 6, normally we need to select 3 more links to form a partially minimum shortest connected graph. These three links are: {(1,2), (4,7), (5,10)}. However, 3 links {(6,7), (6,8) and (7,8)} in the specified set form a loop, and therefore the 9 links do not form a connected graph. The next link to be selected is (7,9) forming PMCG, with a loop. It is shown in Figure 4.7.

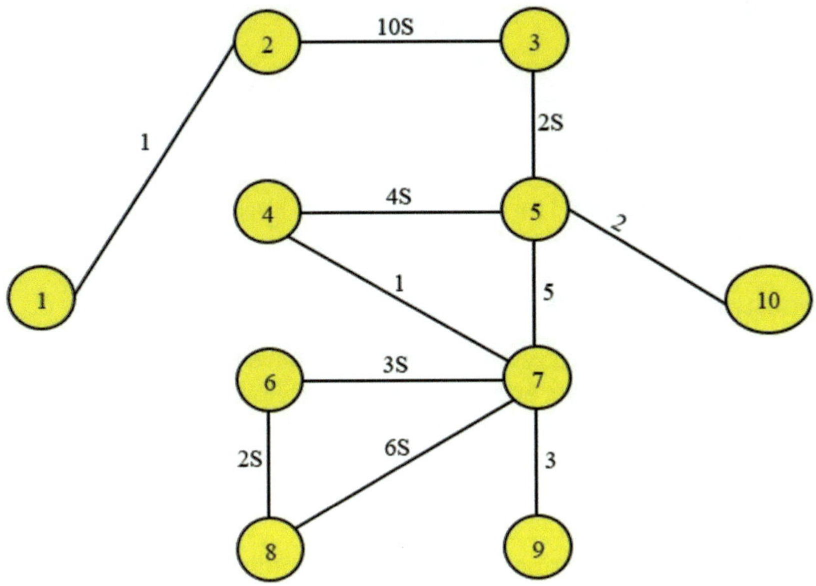

Figure 4.7 PMST of the graph in Figure 4.5.

Note all index orders are balanced, except node 9 and node 7, with index value 1 and 5. Since the link (7,9) can be removed, and it will bring index of all nodes in balance and hence the path can be traced as follows:

$1 \to 2 \to 3 \to 5 \to 4 \to 7 \to 8 \to 6 \to 7 \to 5 \to 10$, with a loop $6 \to 7 \to 8 \to 6$. Total cost will be $= 1+10+2+4+1+3+2+6+5+2 = 36$ units. The required path is shown in Figure 4.8.

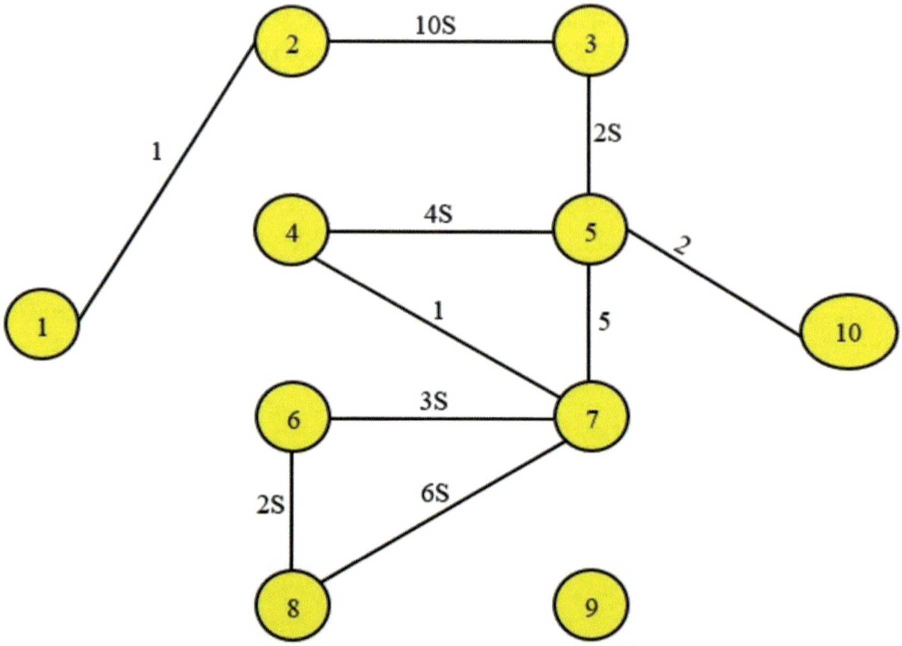

Figure 4.8 Path through the specified links.

4.7 Concluding Remarks

In this chapter, we considered two variants of the conventional routing problem, where a path between the origin and the destination node is constrained to pass through '*K*' specified node or '*K*' specified links. A similar problem through *k* specified nodes was attempted by Saksena and Kumar (1966) and path through '*K*' specified links was also considered by Arora and Kumar (1993) using the ring-sum approach.

It is easy to see that the approach discussed in this chapter, can easily establish lower and upper bound on path length. For example, the lower bound on the path length is the unconstrained path and upper bound is the length of path one gets by the approach discussed in this chapter.

It is desirable to establish a procedure of moving the bounds and thus establishing optimality of the solution.

Computational complexity does not increase with increase in the number of specified links.

References

R. K. Ahuja, K. Melhotra, J. Orlin, and R. E. Tarjan, Fast algorithms for shortest route problem, Journal of the Association of Computing Machinery, 37(2), pp213-223, 1990.

H. Arora, and S. Kumar, Path through '*k*' specified edges in a linear graph, Opsearch, 30(2), 163-173, 1993.

R. E. Beckwith, Dynamic Programming and Network Routing: An introduction to the technique of Functional Equations, Proc. Of the dynamic programming workshop, Boulder, Colorado, 1961.

R. E. Bellman, On an optimal routing problem, Q Appl Math, 16, pp87-90, 1956.

R. E. Bellman, Dynamic Programming treatment to the travelling salesman problem, J of ACM, 9(1), pp61-63, 1962.

G. B. Dantzig, Discreate-variable extremum problem, Opns. Res., 5(2), pp266-277, 1957.

R. Kalaba, On some communication routing problems, Combinatorial Analysis, Proc. Symp. App. Maths. 10, 1960.

S. Kumar, Optimization of Protean Systems – A review, APORS 1994, M. Fushimi and K. Tone (Editors) World Scientific Publishers, pp 139-146, 1995.

S. Kumar, H. Arora, C. Chiera, and G. Raviganesh, Paths for self-healing Protean Networks, Bulletin of the Allahabad Mathematical Society, Vol 14, 1999a.

S. Kumar, G Raviganesh, J. McPhee, T. M. Nyathi, Terminal pair reliability in a Protean network, International Journal of Operations and Quantitative Management, 5(2), pp133-150, 1999b.

S. Kumar, Information recycling mathematical methods for protean systems, A path-way approach, South African Journal of Industrial Engineering, 16(2), pp81-101, 2005.

S. Kumar, Information recycling mathematical methods for Protean Systems: A Path-way approach to Geometric Program, South African Journal of Industrial Engineering, 17(2), pp 127-143, 2006.

S. Kumar, E. Munapo, O. Ncube, C. Sigauke, P. Nyamgure, A minimum weight labelling method for determination of the shortest route in a non-directed network, Int. J. Syst. Assur. Eng. Manag. DOI 10.1007/s198-012-0140-7, 2013.

S. Kumar, E. Munapo, M. Leasoana, and P. Nyamgure, A minimum spanning tree approximation to the routing problem through 'k' specified nodes, J Economics, 5(3), 307-312, 2014.

S. Kumar, E. Munapo, M. Leasoana, and P. Nyamgure, Is the travelling salesman problem NP Hard? Chapter 3 in Engineering and Technology: Recent Innovations and Research, Ed. A Mathani, International Research Publication House, ISBN 978-93-86138-06-4, pp37-58, 2016.

S. Kumar, E. Munapo, M Leasoana, and P. Nyamgure, A minimum spanning tree based for the travelling salesman tour, Opsearch, DOI 10.1007/s1259-017-0318-5, pp1-15, 2017.

S. Kumar, E. Munapo, C. Siguake, and M. Al-Rabeeah, the minimum spanning tree with index <=2 is equivalent to the travelling salesman tour, Chapter 8 in Mathematics in Engineering Sciences: Novel Theories, Technologies and Applications, Ed Mange Ram, CRC Press, ISBN 13:978-1-138-57767-1, 2020.

S. Kumar, E. Munapo, P. Nyamgure, and T. Tawanda, Minimum spanning tree approximation to a path through K specified links, Journal of Graphic Era University, Vol 11, No 2, pp 221-238, 2023, doi: 10.13052/jgeu0975-1416.1127@2023 River Publishers

J. B. Kruskal, On the shortest spanning tree of a graph and the travelling salesman problem, Proc. Amer. Math. Soc., 7, pp48-50, 1956.

E. Munapo, and S. Kumar, Linear Integer Programming: Theory, Applications and Recent Developments, De Gruyter Series on the Applications of Mathematics in Engineering and Information Sciences, ISBN 978-3-11-070292-7, 2022.

R. M. Peart, P. H. Randolph, T. E. Bartlet, The shortest route problem, Opns Res. 8(6), pp866-868, 1960.

M. Pollack, W. Wiebenson, Solution of the shortest route problem: A review, Opns. Res, 8(2), pp224-230, 1960.

A. Rais, F. Alvelos, M. S. Carvalho, New mixed integer programming model for the pickup-and-delivery problem with trans-shipment, European Journal of Operations Research, 235(3), pp530-539, 2014.

J. P. Saksena, and S. Kumar, The routing problem through 'k' specified nodes, Opns. Res. 14(5), 909-913, 1966.

F. B. Zhan, Three Fastest shortest path algorithms on real road networks: Data structured procedures, Journal of the Geographic Information and decision analysis, 1(1), pp70-82, 1997.

5

Is Determination of the Travelling Salesman Tour an NP Hard Problem? Some Polynomial Time Reconstruction Approaches

Abstract

The minimum travelling salesman problem has attracted lot of attention as it has many industrial applications and computationally, it has been classified as NP hard. Algorithms have been developed and many situations have been successfully resolved. In this Chapter, we establish that at least in some cases, the minimum travelling salesman tour is not an NP hard problem. This conclusion is based on two different approaches, discussed in this chapter. First approach involves developing a binary variable mathematical model for the travelling salesman problem, adding some sub-tour elimination constraints, which are also functions of binary variables, and finally transforming that model as a quadratic programming program and solve that model by the interior point linear programming approach. The second approach uses the minimum spanning tree of a given network and converts that spanning tree into a node index restricted spanning tree and finally obtain the minimum salesman tour. In this approach, finally, we establish that the index restricted spanning tree is equivalent to the minimum travelling salesman tour.

Keywords: Binary variable model, quadratic programming model, minimum spanning tree under index restriction, NP hard, polynomial time approach to salesman tour, network reconstruction approach, sub-tour elimination constraints.

Notation

1. $G(n, m)$ denotes a connected network with nodes given by the set $n = \{1,2,\ldots,N\}$ and links are denoted by the set $m = \{1,2,\ldots,L\}$ where each link is joining some nodes i and j, where $i, j = 1, 2, \ldots, n$ and $i \neq j$.
2. $G'(n', m')$ denotes a reconstructed network, where once again the node set is represented by the set n' and the link set is represented by the set m', where the sets n', m' are some reconstructed forms of the given sets n, m, respectively.

5.1 Introduction

The travelling salesman is a classical problem in graph theory, which deals with the determination of the salesman tour starting from the home city, visiting each city once and returning to the home city. Such tours are required in resolving many industrial situations and due to this practical importance, the problem has attracted a lot of attention from researchers and many solution approaches have been suggested from time to time. Researchers continue to have interest in developing new and computationally more effective approaches for this combinatorial problem. Due to computational difficulties, this problem has been classified as NP hard problem, see Nemhauser and Wolsey (1989). However, in this chapter, we raise a question, and to ask is the travelling salesman problem really a NP hard problem? At least in some cases, we present some illustrations that cannot be classified as non-polynomial for the travelling salesman tour. These cases are:

1. The first approach discussed in this chapter is based on a transformation that converts a binary variable problem to a quadratic programming problem and by changing binary variables into bounded variables. This was developed by Munapo (2020). Details of this approach have been presented in Section 5.2.
2. The second approach is based on the index restricted minimum spanning tree that was developed by Munapo et al. (2016). Kumar et al. (2018, 2016), used the idea index restricted minimum spanning tree to develop the salesman tour, which is discussed in Section 5.3. In these approaches, computational requirements do not explode as the number of nodes increases in the given connected network.

In Section 5.4 we develop a reconstruction procedure and establish that the spanning tee under the index restriction gives the minimum spanning tree, which is also the travelling salesman tour, thus establishing that the minimum travelling salesman tour is equivalent to the minimum spanning tree under a node index restriction. This approach was discussed by Kumar et al. (2019).

Finally, the chapter is concluded in Section 5.5.

TSP has been discussed by many researchers, for example, see Finke et al. (1983), Gavish and Graves (1978), Lenstra and Rinnooykan (1975), Miller (1960), Schrijver (1998), Vajda (1961), and Wong (1980). These references are not intended to be exhaustive.

5.2 Binary Formulation of the Travelling Salesman Problem

Munapo (2020) developed a mathematical programming model for the travelling salesman problem based on 5 structural components and transformed that model into a quadratic programming model and solved it as a quadratic programming program using the interior point approach. His considerations included:

Consideration 1: In a travelling salesman tour, if we are at some node k, from that node *k*, if the number of links emanating is *r*, as shown in Figure 5.1, we will use two of these links, one link to reach the node *k* and again use a different link to come out from the node *k*. This requirement will give rise to a constraint (5.1), if variables representing these r links are given by $x_{k1}, x_{k2}, \ldots, x_{kr}$, the constraint with respect to the node *k* will be:

$$x_{k1} + x_{k2} + \cdots + x_{kr} = 2. \tag{5.1}$$

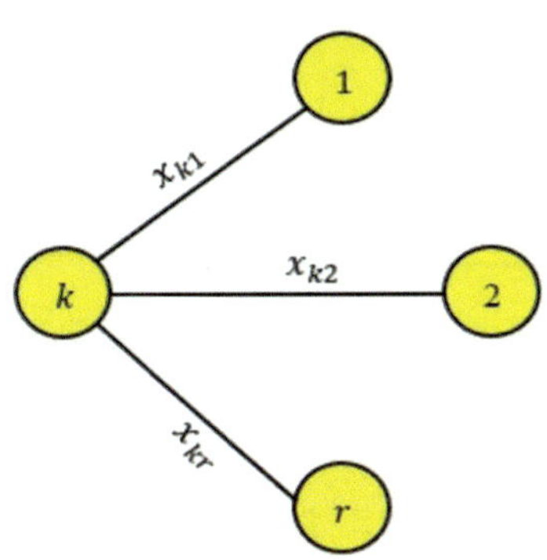

Figure 5.1 A typical node k, from where r number of links are emanating.

In the constraint (5.1), all are binary variables, i.e., $x_{k1}, x_{k2}, \ldots, x_{kr}$ are 0 or 1, 0 indicating that link is not included in the travelling salesman tour and 1 indicates that the link is a part of the salesman tour. Note that we need one constraint for each node, $k = 1, 2, \ldots, N$. Therefore, there will be *N* number of such constraints, one for each node in the given network.

Constraint (5.1) has limitations, as it can give rise to sub-tours. More details can be seen in Papadimitriou and Steiglitz (1998). The subtours can be eliminated by adding some additional constraints, as discussed in Consideration 2.

Consideration 2: Sub-tours can be eliminated by adding sub-tour elimination constraints. Suppose in a 9-node network, we have two sub-tours made by nodes {1,2,3,4} and {5,6,7,8,9} as shown in Figure 5.2. This 9-node network has 9 links forming two sub-tours as shown in Figure 5.2. All links that join a node from sub-tour 1 to a node in sub-tour 2, form a link that will be included in the sub-tour elimination constraints. For example, we illustrate, in Figure 5.2. Five such links, shown in Figure 5.2 join a node from the sub-tour 1 to a node in sub-tour 2. These links are {(2,7), (3,8), (1,6), (4,5), (4,9)}. The sub-tour elimination constraint will be as given by eqn (5.2).

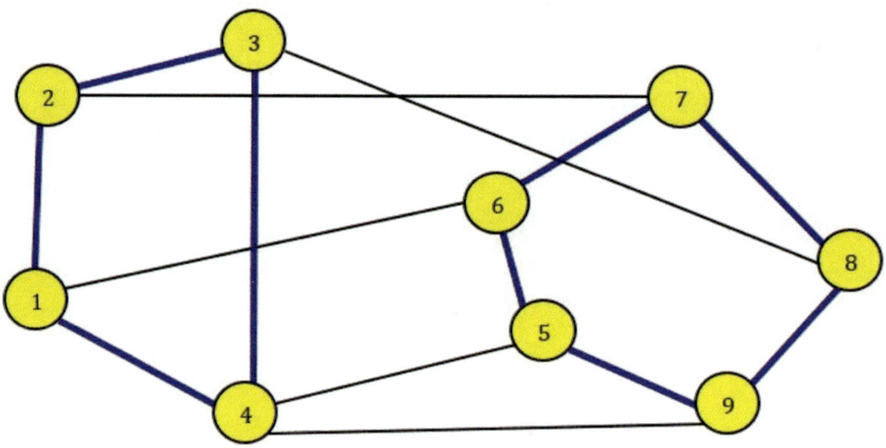

Figure 5.2 Two sub-tours made by nodes {1,2,3,4} and {5,6,7,8,9}.

$$x_{16} + x_{27} + x_{38} + x_{45} + x_{49} \geq 2. \tag{5.2}$$

Sub-tours arise when at least three nodes are joined by three links. Alternatively, these links can also be identified as they form a cut between the two sub-tours. If we can identify a cut after every set of three nodes, a sub-tour elimination constraint can be developed. It is demonstrated in dealing with a numerical illustration later in this section.

Consideration 3: Objective of the travelling salesman tour is to minimize the total cost of the tour. This is reflected by objective function in eqn (5.3)

$$\text{Minimize } Z = c_{12}x_{12} + c_{13}x_{13} + \cdots + c_{1r}x_{1r} + \cdots + c_{kn}x_{kn}. \tag{5.3}$$

Note the number of terms in the objective function will be equal to the number of links in the given network. In a completely connected 'n' node network, the number of links will be given by $\{n(n - 1)/2\}$.

Consideration 4: Non-negative conditions on the variables, as mentioned under the consideration 3, each link in the given network represents a variable, and these variables are binary, i.e.,

$$x_{ij} = 0 \text{ or } 1,\ i, j = 1, 2, \ldots,\ n; i \neq j \tag{5.4}$$

Zero indicated non-inclusion of a link in the minimum salesman tour and one indicates inclusion of the link. However, with the help of a transformation, Munapo (2020) converted them as bounded variables, i.e., he considered them as given by eqn (5.5). They are:

$$0 \leq x_{ij} \leq 1 \text{ for each link}(i, j) \tag{5.5}$$

Consideration 5: Total number of links in the selected salesman tour in an 'n' node network will be 'n', this results in a constraint in eqn (5.6):

$$x_{12} + x_{13} + \cdots + x_{1r} + \cdots + x_{kn} = n \tag{5.6}$$

The above five considerations lead to a mathematical model given by eqn (5.7)

$$\text{Minimize} \quad Z = c_{12}x_{12} + c_{13}x_{13} + \cdots + c_{1r}x_{1r} + \cdots + c_{kn}x_{kn}.$$

Subject to:

$$x_{k1} + x_{k2} + \cdots + x_{krk} = 2,$$

rk is the number of emanatting links from the node k, $k = 1, 2, \ldots, n$.

Total number of selected links:

$$x_{12} + x_{13} + \cdots + x_{1r} + \cdots + x_{kn} = n.$$

Subtour elimination constraints like consideration 2, can be developed along the *x* and *y* directions of the given network, after three nodes, draw a cut and include that as a subtour elimination constraint. Number of such constraints will depend on the structure of the given network.

Finally, replace the binary variables by the bounding constraints:

$$0 \leq x_{ij} \leq 1, \ \forall \, (i, j) \, . \tag{5.7}$$

In the above model, if we return to binary constraints on each variable, the integer programming does not have any polynomial time approach to solve it; however, Munapo (2016) transformed this binary model to bounded variable constraints as shown in eqn (5.7) and changed the problem to a quadratic programming model given by eqn (5.8).

$$\text{Min} \quad f\left(\overline{X}\right) = l_1 CXX^T + SS^T + l_2 XS^T.$$

Subject to:

$$\begin{aligned} AX^T &\geq B^T \\ X^T + S^T &= 1^T. \end{aligned} \tag{5.8}$$

Note that models (5.7) and (5.8) are equivalent; therefore, the solution of eqn (5.8) gives solution that also satisfies eqn (5.7). Their equivalence is established by the following relations:

$$\sum_{i=1}^{n} C_i x_i = \sum_{i=1}^{n} C_i x_i^2 \tag{5.9}$$

$$\sum_{i=1}^{n} s_i = \sum_{i=1}^{n} s_i^2. \tag{5.10}$$

Note that eqn (5.9) and (5.10) are equal at binary values of x_i and s_i, i.e., 0 or 1.

The last component of the objective function has a constant l_2, which is a large constant and when the variables assume binary values, each term is equal to zero as when $x_i = 1$, s_i will be 0 and similarly $x_i = 0$, s_i will be 1 but the $x_i.s_i = 0$ for each i. Any non-binary value will make this term non-zero, and Munapo (2020) calls this term 'enforcer', i.e., real variables are forced to take binary values. This is a crucial point, which changes the binary integer programming model to a quadratic programming model, where the unknowns become bounded real variables, and integer programming model can be solved as a quadratic programming model with upper bounded variables. This transformation eliminates solving the problem as a binary integer variables problem.

5.2.1 Steps of the Munapo (2020) algorithm

Step 1: From the given network diagram, form a binary integer program for the travelling salesman tour.

Step 2: From the given network diagram, develop sub-tour elimination constraints and add these constraints to the model obtained in Step 1.

Step 3: Convert the combined binary integer program into a quadratic program with bounding constraints.

Step 4: Solve the convex quadratic programming problem using the interior point algorithm.

5.2.2 A numerical illustration

Let us consider the example considered by Kumar et al. (2019) given below in Figure 5.3 and link weights are given in Table 5.1.

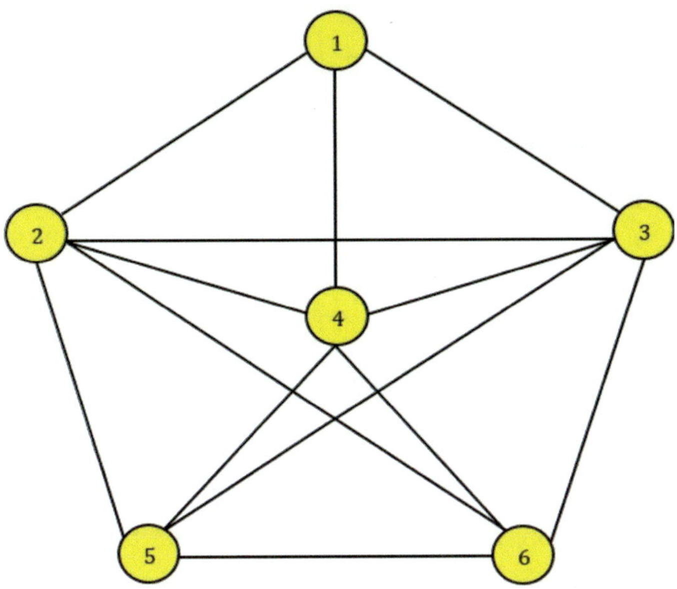

Figure 5.3 Network for illustration of Munapo's approach to find the TSP.

Table 5.1 Link weights.

From/to	1	2	3	4	5	6
1	-	11	9	9	-	-
2	11	-	14	10	10	15
3	9	14	-	6	13	11
4	9	10	6	-	9	10
5	-	10	13	6	-	8
6	-	15	11	10	8	-

The mathematical model will be given by eqn (5.11) with several components as follows:

$$\begin{aligned}\text{Min } Z &= 11x_{12} + 9x_{13} + 9x_{14} + 14x_{23} + 10x_{24} + 10x_{25} + 15x_{26} \\ &\quad +6x_{34} + 13x_{35} + 11x_{36} + 9x_{45} + 10x_{46} + 8x_{56}\end{aligned} \tag{5.11-1}$$

Subject to:

(1) Constraints with respect to each node

$$\begin{aligned}\text{Node1}: &\quad x_{12} + x_{13} + x_{14} = 2 \\ \text{Node 2}: &\quad x_{12} + x_{23} + x_{24} + x_{25} + x_{26} = 2 \\ \text{Node 3}: &\quad x_{13} + x_{23} + x_{34} + x_{35} + x_{36} = 2 \\ \text{Node 4}: &\quad x_{14} + x_{24} + x_{34} + x_{45} + x_{46} = 2 \\ \text{Node 5}: &\quad x_{25} + x_{35} + x_{45} + x_{56} = 2 \\ \text{Node 6}: &\quad x_{26} + x_{36} + x_{46} + x_{56} = 2.\end{aligned} \tag{5.11-2}$$

(2) Sub-tour elimination constraints: Since the network is comprised of 6 nodes, only one constraint in direction x and one constraint in the y direction will be sufficient. See Figures 5.4 and 5.5.

For the network in Figure 5.3, the constraint will be developed from the links shown in Figures 5.4 and 5.5 is given below.

From Figure 5.4, we obtain a sub-tour elimination constraint comprised of 8 elements, given by eqn (5.11-3a).

$$x_{13} + x_{14} + x_{23} + x_{24} + x_{26} + x_{54} + x_{53} + x_{56} \geq 2. \tag{5.11-3a}$$

Similarly, from Figure 5.5, the constraint will again be of 7 elements, given by eqn (5.11-3b).

$$x_{25} + x_{26} + x_{24} + x_{14} + x_{34} + x_{35} + x_{36} \geq 2. \tag{5.11-3b}$$

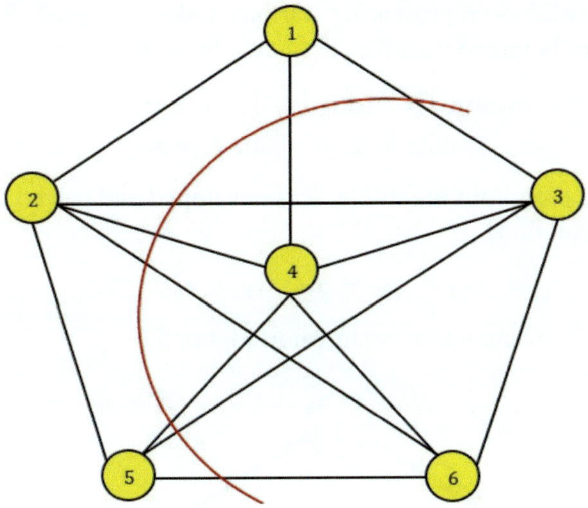

Figure 5.4 Cut in the x-direction for the sub-tour elimination constraint.

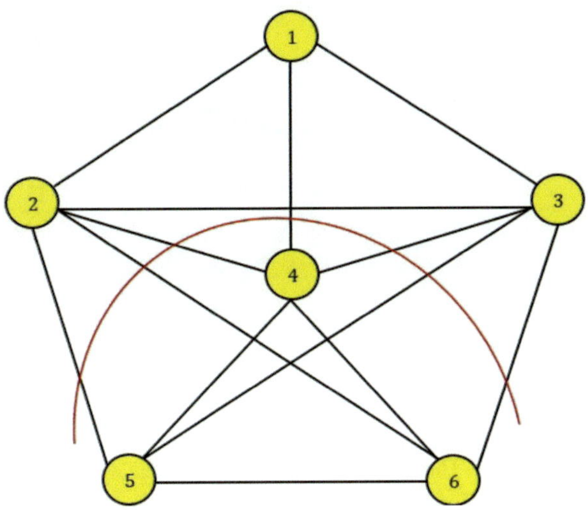

Figure 5.5 Cut in the y-direction for the sub-tour elimination constraint.

Total number of selected links will be 6 in a six-node network. This constraint will be eqn (5.11-4).

$$x_{12} + x_{13} + x_{14} + x_{23} + x_{24} + x_{25} + x_{26} + x_{34} + x_{35} + x_{36} + x_{45} + x_{46} + x_{56} = 6. \qquad (5.11\text{-}4)$$

Finally, all variables representing various links in Figure 5.3 will be binary variables to bounded variables as shown by eqn (5.11-5):

$$x_{ij} \text{ are } 0 \text{ or } 1 \text{ for } i = 1,\ 2,\ 3,\ 4,\ 5 \quad \text{and } j = 2,\ 3,\ 4,\ 5,\ 6,\ i \neq j. \tag{5.11-5}$$

The quadratic programming model was developed and solved. It resulted in the following solution.

$$x_{12} = x_{13} = x_{25} = x_{34} =_{46}= x_{56} = 1. \tag{5.12}$$

The minimum salesman tour is shown in Figure 5.6.

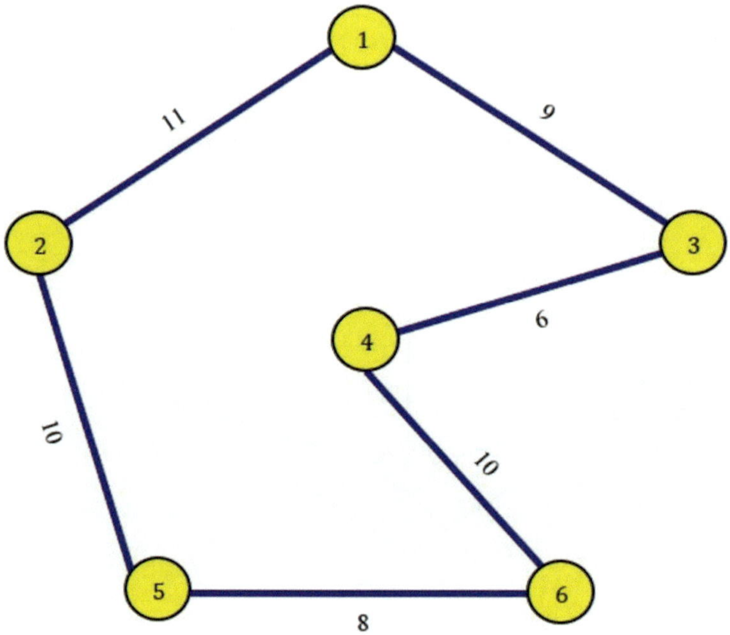

Figure 5.6 The minimum travelling salesman tour.

Total minimum cost will be given by $11 + 9 + 10 + 6 + 10 + 8 = 54$.

5.3 The Travelling Salesman Tour Based on the Index Restricted Minimum Spanning Tree

Kumar et al. (2018) considered a connected network, $G(n, m)$, where n denotes the number of nodes and m denotes the number of links in the network $G(n,m)$. From the network $G(n,m)$, they constructed another network $G'((n-1),(m-d_p))$, where $(n-1)$ denotes the number of nodes in the reconstructed network and $(m- \ d_p)$ denotes the number of links in the same network. Here d_p denotes the number of links emanating from the node p, where the node p is one of the given nodes of the network $G(n,m)$. Usually, the node p is selected as the node with a minimum number of links emanating from it. If there are more than one node with an equal number of links emanating from them, the node p can be selected arbitrarily. Since the minimum travelling salesman tour must pass through each link, the selected node p can be any node of the given network $G(n,m)$.

Note that the networks $G(n,m)$ and $G'((n-1),(m-d_p))$ are essentially similar except that the network $G'((n-1),(m-d_p))$ is a reconstructed network with one node less, i.e. from the n nodes, the node p is missing in the reconstructed network. Similarly, from the links, the d_p number of links are removed, which originated from the node p. From the deleted d_p number of links, select two links (pq) and (pk) and find an index restricted minimum spanning tree in the network $G'((n-1),(m-d_p))$ joining the nodes (q) and (k), i.e., the index at the nodes q and k is one and for all other nodes the index will be 2. That index restricted path together with the links (pq) and (pk) will form a travelling salesman tour, hence its value will form an upper bond on the required minimum salesman tour. Similarly, one can find a salesman tour for the other combinations, and the number of these combinations will be given by d_{pC_2}. Each combination will rise to a salesman tour and minimum of them will the required minimum salesman tour. Here we will be using ideas from the Chapter 3, where we obtained a minimum spanning tree under the index restriction at each node and will also make use of the index balancing theorem discussed in chapter 3.

5.3.1 Algorithm to find the travelling salesman tour

The algorithm is composed of the following steps:

Step 1: From the given network $G(n,m)$, identify and carry out actions as described under:

1.1 Identify the node p, with minimum number of links emanating from it.

1.2 Find the index of the node p, i.e., the number of links emanating from this node. Let this number be represented by r, where $r \geq 2$.

1.3 Make possible combinations, taking two at a time from these r number of links that can be made. Let this number be represented by l, it is given by $r_{C2} = 1, 2, \ldots, l$. Let us denote these combinations by $C_1, C_2, \ldots, C_l$ and their corresponding lengths of two links, be denoted by $L(C_1), L(C_2), \ldots, L(C_l)$ such that $L(C_1) \le L(C_2) \le \cdots \le L(C_l)$.

1.4 Identify the nodes q and k associated with the minimum length $L(C_1)$.

1.5 Reconstruct the network:

$$G'((n-p), (m-r)) \text{ and set the length of the link } (q,k) = \infty.$$

1.6 Find the minimum spanning tree (MST) of the reconstructed network,

$$G'((n-1), (m-r)), \text{where the link } (q,k) = \infty.$$

1.7 Find the length of the MST obtained in 1.6 above and denote this by: LMST$G'((n-1), (m-r))$. Find the number of index-imbalances in the MST $G'((n-1), (m-r))$ and let this number be denoted by $NoMSTG'((n-1), (m-r))$.

1.8 Apply the index balancing theorem from Chapter 3, until the MST becomes an MST path joining the nodes q and k for the combination C_1.

1.9 Develop a feasible travelling salesman tour and denote its length by $LTSTG(n,m) = L(p,q) + L(p,k) + LMSTPG'((n-1)(m-r))$.

1.10 Since this is a feasible travelling salesman tour, it becomes an upper bound for the required minimum salesman tour. It will be given by the length obtained in 1.10 above.

1.11 Set a counter $K = 1$ and go to Step 2.

Step 2: Set $K = K + 1$. If $(K + 1) > l$, go to Step 4. If $(K + 1) < l$, return to Step 3 and carry out the analysis to find the minimum salesman tour with respect to the next combination.

Step 3:

3.1 Identify the two associated nodes with the $(K + 1)$th combination. Once again find the minimum salesman tour, as was determined earlier in Step 1 for the Kth combination.

3.2 If the salesman tour is less than the current upper bound, replace the upper bound with the current value, else once again return to Step 2.

Step 4: The minimum salesman tour has been identified and its value is the current upper bound.

5.3.2 Numerical illustration

Consider the network $G(8, 14)$ as shown in Figure 5.7 and link lengths are given in Table 5.2, which was considered by Kumar et al. (2018).

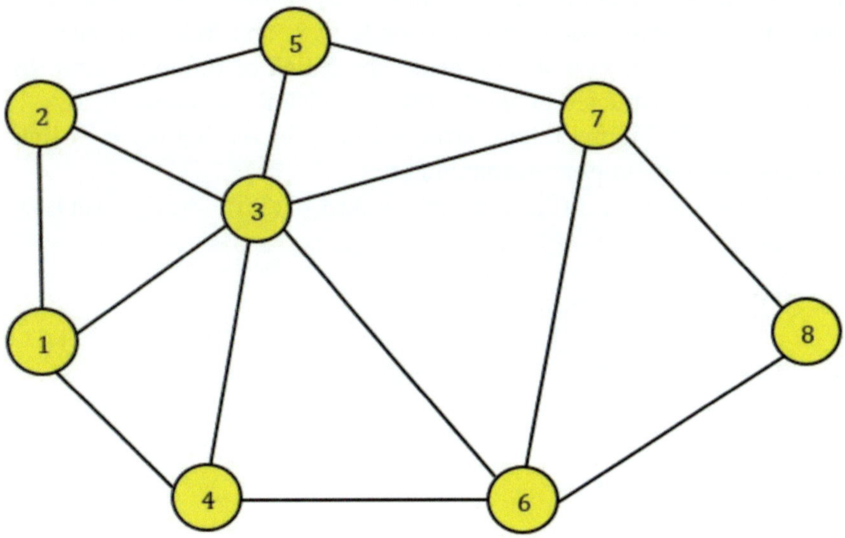

Figure 5.7 A network for finding the minimum travelling salesman tour.

Table 5.2 Link weights for the links in Figure 5.7.

From/to	1	2	3	4	5	6	7	8
1	-	6	4	10	∞	∞	∞	∞
2	6	-	6	∞	6	∞	∞	∞
3	4	6	-	4	7	6	14	∞
4	10	∞	4	-	∞	12	∞	∞
5	∞	6	7	∞	-	∞	8	∞
6	∞	∞	6	12	∞	-	4	2
7	∞	∞	14	∞	8	4	-	3
8	∞	∞	∞	∞	∞	2	3	-

The number of emanating links from each node, i.e., the index for each node in Figure 5.7 is given in Table 5.3.

Table 5.3 Index of each node in the given network.

Node	1	2	3	4	5	6	7	8
Index	3	3	6	3	3	4	4	2

The selected node p in this case will be node 8, as it has minimum number of links emanating from it. There is only one combination, hence the salesman tour will be the required minimum salesman tour. The links joining the node 8 are (7,8) and (6,8). The reconstructed network is as shown in Figure 5.8, and it is given by $G\left((8-1),(14-2-1)\right)$ or $G\left(7,11\right)$ below. Note that node 8 will be removed, two links joining this node (7,8) and (6,8) will also be removed and the link (6,7) will be replaced by a link of infinite length. In other words, the reconstructed network will have 11 links as shown below.

These 11 links in Figure 5.8 are arranged in an increasing order of link lengths to construct the minimum spanning tree.

4(1,3), 4(3,4), 6(1,2), 6(2,3), 6(2,5), 6(3,6), 7(3,5), 8(5,7), 10(1,4), 12(4,6), 14(3,7).

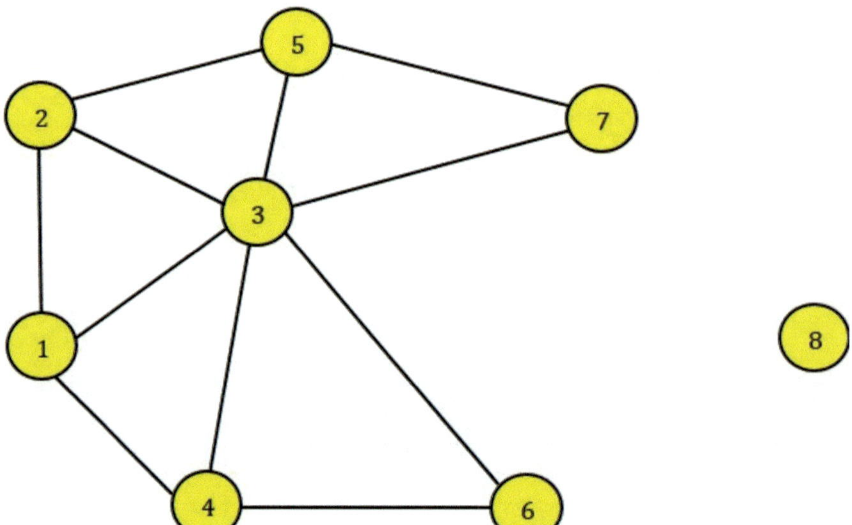

Figure 5.8 The reconstructed network from the network in Figure 5.7.

The minimum spanning tree of the network in Figure 5.8 is given in Figure 5.9.

Note that the imbalance with respect to the index restricted spanning tree is given in Table 5.4.

One application of the index balancing theorem from Chapter 3 will be sufficient for this network. However, there are two ways to attain the balance,

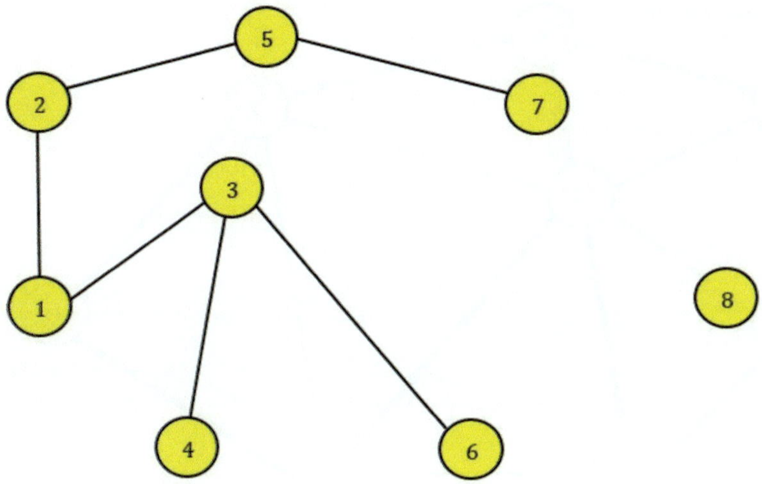

Figure 5.9 The minimum spanning tree of the network in Figure 5.8.

Table 5.4 Index imbalances for the minimum spanning path joining nodes 6 and 7.

Node	**1**	**2**	**3**	**4**	**5**	**6**	**7**
Index	2	2	3	1	2	1	1
Desired	2	2	2	2	2	1	1
Imbalance	-	-	-1	+1	-	-	-

i.e., remove one link emanating from node 3 and add a link to join node 4. This can be achieved in two different ways:

1. Remove link (1,3) and add link (1,4), total increase will be $(10 - 4) = 6$ or alternatively, the second possibility will be:
2. Remove the link (3,6) and add the link (4,6), total increase is once again $(12 - 6) = 6$.

These two minimum spanning tree paths together with the links (6,8) and (7,8) will form the minimum spanning trees, which are shown in Figures 5.10 and 5.11.

$$Z = 45,\ x_{12} = x_{14} = x_{43} = x_{25} = x_{36} = x_{57} = x_{68} = x_{78} = 1.$$

The alternative answer is shown in Figure 5.11.

$$Z = 45,\ x_{12} = x_{13} = x_{43} = x_{25} = x_{46} = x_{57} = x_{68} = x_{78} = 1.$$

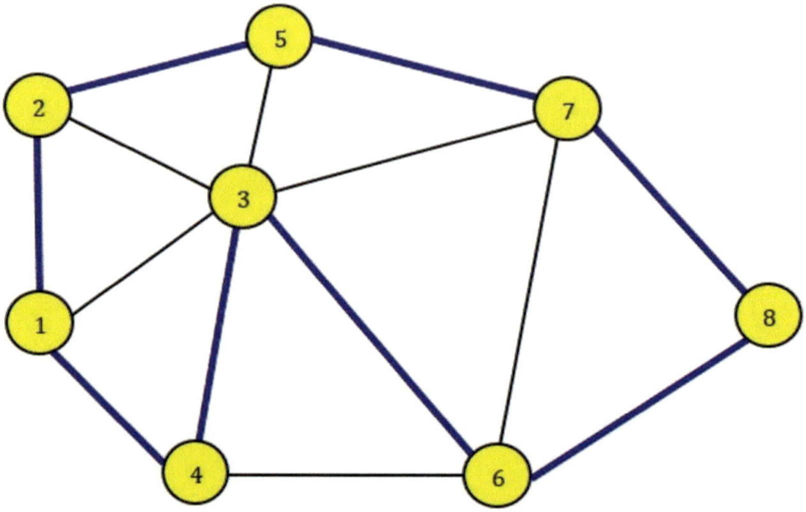

Figure 5.10 The travelling salesman tour is shown in blue lines.

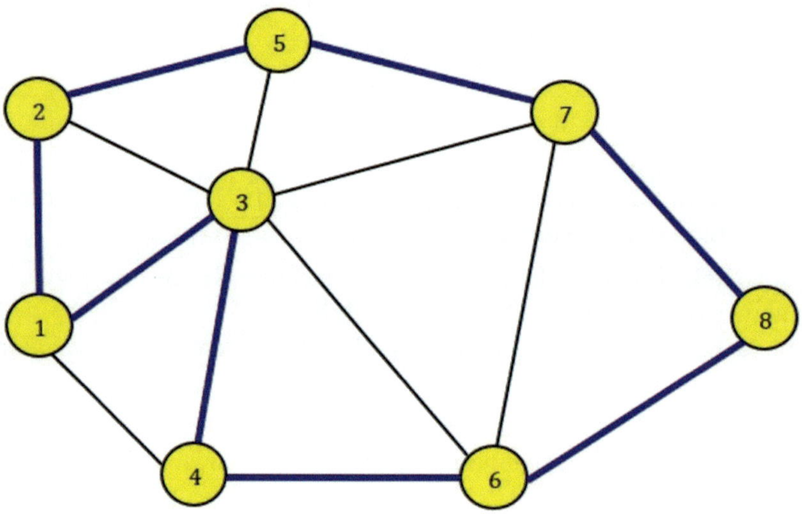

Figure 5.11 The alternative minimum spanning tree.

5.3.3 Remarks and observations

In the above numerical illustration, we had a node that was having only two links emanating from it. Since, in the travelling salesman tour, each node is visited, we need at least two links, one for entry to that node and another for an exit from that node. Hence, in this example, we had only one combination of taking two from the two. These combinations quickly start increasing as index of the minimum index node increase. For example, number of combinations grows, as shown in Table 5.5.

Table 5.5 Number of combinations as a function of the index value of a node.

Node index	2	3	4	5	6
Number of Combinations	1	3	6	10	15

From the above table, one can easily see that computational load will increase for obtaining the minimum travelling salesman tour. This computational load was reduced by taking one link at a time, for example, with 6 as the index value, one must solve 15 problems, one for each combination and if we take one link at a time, we can get a similar outcome by solving only 6 problems. This was discussed by Kumar et al. (2016).

5.4 Node Index Restricted Spanning Tree Is Equivalent to the Minimum Travelling Salesman Tour

5.4.1 A new approach to the index restricted minimum spanning tree and some implications

In Section 5.3, we first selected arbitrarily two links (p, q) and (p, k) from a selected node p and reconstructed a new network without the node p, and without all the links emanating from the node p. In this reconstructed network, we obtained an index restricted minimum spanning tree joining the nodes q and k, which made a feasible travelling salesman tour in the given network $G(n, m)$ when links (p, q) and (p, k) were added to this minimum spanning path joining the nodes q and k. Idea in the next section is to avoid the arbitrary selection of two links from the node p. In the approach, we are going to discuss in this section, our aim is that we let the optimization approach make its own selection of the links emanating from the node p. We achieve this by making the node p duplicate and all links emanating from the node p are also added to its duplicate node p'. Furthermore, if we assume that the node p and its duplicate p' are joined by a link with link length zero, then the index restricted minimum spanning tree joining the nodes p and the node p' will form a travelling salesman tour. However, the duplicate links will require some special considerations, which are resolved in Section 5.4.2. Due to these duplicate links, there will be some changes in conventional approach for selecting a link in the minimum spanning tree, which will be discussed in Section 5.4.2.

5.4.2 Network reconstruction

Here the network $G(n, m)$ can be modified in a slightly different way. In Section 5.3, we considered $G'(n', m')$, where we removed some nodes and links; however in this section, instead of removing a node and some links in the reconstructed network, we consider adding nodes and some links. From the given network, we develop a reconstructed network $G'((n+1), (m+d_p))$, adding a node and links equal to the number of links emanating from the node p. We explain this by an illustration. Consider Figure 5.12 as the given network $G(n, m) = G(4, 6)$.

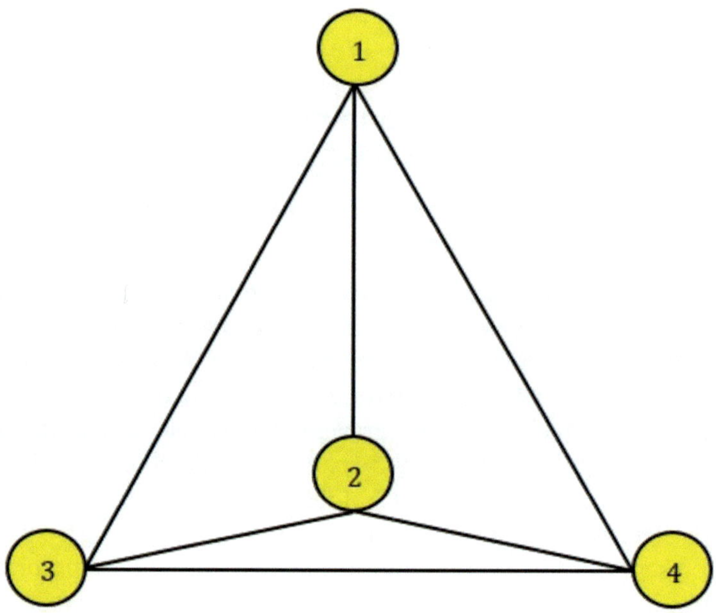

Figure 5.12 The network $G(4, 6)$.

From every node in Figure 5.12, the number of links emanating are equal to 3; hence any node may be selected to serve as the node p, and the value of $d_p = 3$ for all nodes in this network. Consider that we select node 1 as the node p, then the network

$G'((n+1), (m+d_p)) = G(5, 9)$, which is shown in Figure 5.13.

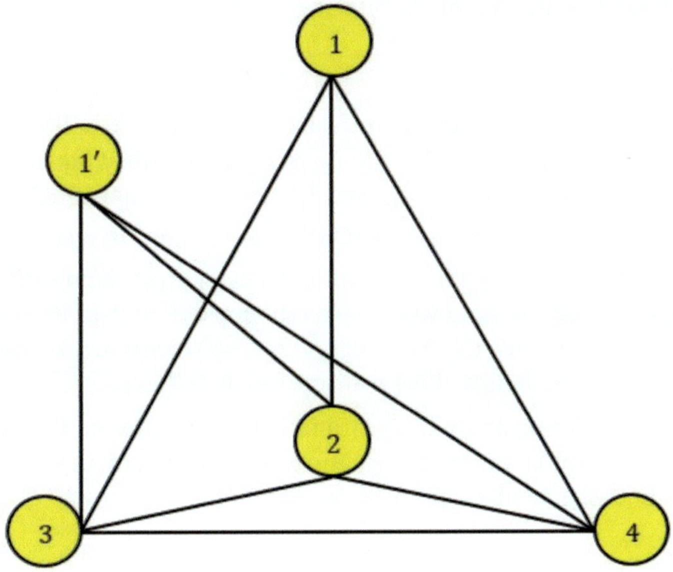

Figure 5.13 The network $G(5,9)$.

Note that in Figure 5.13, we have added an additional node 1' and three additional links (1',2), (1',3) and (1',4). Note the node 1' is not real and similarly the links (1',2), (1',3) and (1',4) are also not real. They have been added for convenience. When a link is selected for inclusion in the minimum spanning tree, that link is called basic and note that the link (1,2) and (1',2) cannot coexist as basic links, as only one link is true and the other is unreal. Therefore, when one of them is selected as a basic link, the other one will be converted to infinite length by a transformation and therefore the duplicate of that link will not be selected for inclusion in the minimum spanning tree. This can be achieved by defining lengths in some modified way, as discussed later in this section.

The basic idea is to merge the duplicate node, which is 1 in the above Figure 5.13. Note that for the network in Figure 5.13, the index restricted minimum spanning tree joining nodes 1 and 1' will be comprised of four links and when the nodes 1 and 1' are merged, it will become a travelling salesman tour. Since we are trying to find a minimum spanning path joining the nodes 1 and 1', it will become the minimum travelling salesman tour. To understand this logic, let us assign some numerical values to various links in Figure 5.12 and see how we assign link lengths to various links in Figure 5.13.

Let us assume these values are given by eqn (5.13):

$$l_{12} = 3,\ l_{13} = 5,\ l_{14} = 6,\ l_{23} = 4, l_{24} = 6, l_{34} = 5. \tag{5.13}$$

The links in Figure 5.13, i.e., the reconstructed network will be of two kinds, for example, the links l_{23}, l_{24} and l_{34} are original (they have not been duplicated), however, the links l_{12}, l_{13} and l_{14} have been duplicated. In each pair $(1, k)$ and $\left(1^{'}, k\right)$, $k = 2, 3, 4$. Only one of them is real, the other one is fictitious. To overcome this difficulty, we categorize them differently. Duplicated links will be assigned values as a function of length of its duplicate as shown in eqn (5.14) and (5.15). Since they can't coexist, we impose a condition eqn (5.18), see below. This is discussed as follows:

1. Lengths of the duplicate links as a function of the original lengths will be given by the following transformation given by eqn (5.14), here s_{ij} is just a parameter given by eqn (5.16) and (5.17).

$$l_{1',2} = \frac{l_{12}}{(1 - s_{12})},\ l_{1',3} = \frac{l_{13}}{(1 - s_{13})},\ l_{1',4} = \frac{l_{14}}{(1 - s_{14})}. \tag{5.14}$$

2. Similarly, the lengths of the original links will also be a function of the lengths of the duplicate links, as given by eqn (5.15)

$$l_{1,2} = \frac{l_{1',2}}{(1 - s'_{12})},\ l_{1,3} = \frac{l_{1',3}}{(1 - s'_{13})},\ l_{1,4} = \frac{l_{1',4}}{(1 - s'_{14})}. \tag{5.15}$$

3. Remaining links will not change. The values of other parameters in relations eqn (5.14) and (5.15) will be governed by the values given by eqn (5.16) and (5.17).

$$s_{1k} = \begin{cases} 1, \text{ if the link } (1k) \text{ is basic} \\ \text{It is going to be 0 otherwise} \end{cases} \tag{5.16}$$

$$s_{1',k} = \begin{cases} 1, \text{ if the link } \left(1^{'}, k\right) \text{ is basic} \\ \text{It is going to be 0 otherwise.} \end{cases} \tag{5.17}$$

To avoid existence of a link and its duplicate, we impose the condition eqn (5.18).

$$s_{1k} + s_{1'k} \leq 1 \tag{5.18}$$

It may be noted that if a link from node 1 is selected for inclusion in the minimum spanning tree, it will be either from node 1 or from node 1', but the same link will not be included from nodes 1 and 1'. In view of these transformations, the link lengths for Figure 5.13 are given in Table 5.6. Note that links from the node 1 or 1' have two possible values, depending on their inclusion in the minimum spanning tree.

In Table 5.6, there are two types of links, i.e., those which may change their value and others for which value does not change. Just to distinguish between them, we just call them Type 2 (that may change the associated weight) and Type 1, which do not change. Here the Type 2 links are (1,2), (1,3), (1,4), (1',2), (1',3) and (1',4) and Type 1 links are ((2,3), (2,4), (3,4)).

Table 5.6 Link lengths for Figure 5.13.

From/to	1	1'	2	3	4
1	-	-	3	5	6
1'	-	-	$\frac{3}{(1-s_{12})} = \infty \; or \; 3$	$\frac{5}{(1-s_{12})} = \infty \; or \; 5$	$\frac{6}{(1-s_{12})} = \infty \; or \; 6$
2	3	$\frac{3}{(1-s_{12})} = \infty \; or \; 3$	-	4	6
3	5	$\frac{5}{(1-s_{12})} = \infty \; or \; 5$	4	-	5
4	6	$\frac{6}{(1-s_{12})} = \infty \; or \; 6$	6	5	-

Since link lengths are subject to change, Kruskal's minimum spanning tree approach will have to be modified to accommodate the possibility of a change in link lengths. This is discussed in Section 5.4.3.

5.4.2.1 Slight modification to Kruskal's Greedy approach and its justification

Kruskal (see http://www.youtube.com/watch?v=3rrNH_AizMA) developed a greedy approach, that has been described as follows:

1. Sort all edges in nondecreasing order of their weights.
2. Pick the smallest edge. Check it forms a cycle with the spanning tree formed so far. If a cycle is not formed, include this edge. Else, discard it.
3. Repeat the step #2 until there are $(n - 1)$ edges in the spanning tree.

The above conventional greedy approach does not consider possibility of links changing their lengths, which is the case in the proposed reconstruction. Let us illustrate a possible difficulty by an illustration. Reconsider the network in Figure 5.12, and the associated reconstructed network with initial link length information as given in Table 5.7. Note initially all links are non-basic, as no link has been selected.

Table 5.7 Link lengths for the current illustration for the network in Figure 5.13.

From/to	**1**	**1'**	**2**	**3**	**4**
1	-	-	1	1	1
1'	-	-	1	1	1
2	1	1	-	2	2
3	1	1	2	-	3
4	1	1	2	3	-

Arranging links in nondecreasing order, we have:

$$1(1,2), 1(1,3), 1(1,4), 1(1',2), 1(1',3), \\ 1(1',4), 2(2,3), 2(2,4), 3(3,4). \tag{5.19}$$

Note that first 6 links in eqn (5.19) are Type 2 and last three links are Type 1. The link (1,2) will be selected, and it being Type 2, will change the length of the duplicate link (1',2) to infinity. Selected links are shown in **bold**; the rearranged links will be as shown in eqn (5.20).

$$\mathbf{1(1,2)}, 1(1,3), 1(1,4), 1(1',3), 1(1',4), \\ 2(2,3), 2(2,4), 3(3,4), \infty(1',2). \tag{5.20}$$

Next move will select the link (1,3) and once again the (1',3) will become ∞. This will give rise to eqn (5.21), given below:

$$\mathbf{1(1,2)}, \mathbf{1(1,3)}, 1(1,4), 1(1',4), 2(2,3), \\ 2(2,4), 3(3,4), \infty\left(1',2\right), \infty\left(1',3\right). \tag{5.21}$$

Once again, the next link to selected is 1(1,4), again being Type 2 will change the length of the link (1',4), resulting in the following:

$$\mathbf{1(1,2)}, \mathbf{1(1,3)}, \mathbf{1(1,4)}, 2(2,3), 2(2,4), 3(3,4), \\ \infty\left(1',2\right), \infty\left(1',3\right), \infty\left(1',4\right). \tag{5.21}$$

The above steps have resulted in the spanning tree shown in Figure 5.14.

Hence Kruskal's greedy approach requires some modifications to overcome these difficulties, which is presented in Section 5.4.3.2.

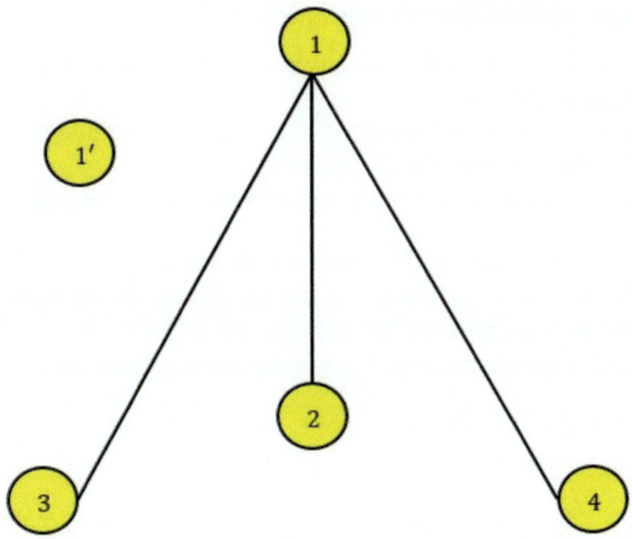

Figure 5.14 An incomplete spanning tree with a disconnected node.

5.4.2.2 Modified steps for obtaining the minimum spanning tree of the reconstructed network $G^{'}\left(\left(n+1\right),\left(m+d_{p}\right)\right)$

Step 1: For the given network $G\left(n,m\right)$, reconstructed the modified network $G^{'}\left(\left(n+1\right),\left(m+d_{p}\right)\right)$, identify the node p and the corresponding d_p and arrange the links in an increasing order. Since at the start, all links are non-basic, hence the lengths of the duplicate links will be the same as given lengths. Total number of Type 2 links will be ($2d_p$) and the remaining ($m - d_p$) Type 1 links, which have a fixed given length. Among Type 2 links, at least two links will have equal length.

Step 2: Set a counter $K = 1$ and select the link of minimum length and include that link in the minimum spanning tree and go to Step 3.

Step 3: If the selected edge was Type 1 and $K < n$, set $K = K + 1$ and select the next minimum length. If $K = n$, go to Step 7.

Step 4: If the selected edge was Type 2 and $K < n$, first set its duplicate length equal to ∞ and rearrange lengths in increasing order. Set $K = K + 1$ and select the next minimum length. If $K = n$, go to Step 7.

Step 5: If the selected link forms a cycle with the already selected links, discard that link else include that link. If it is Type 1, return to Step 3. If the selected link is Type 2, go to Step 6.

Step 6: With respect to the selected link, check

1. That the link does not lead to isolation of a node in the reconstructed network, if so, discard it.
2. Its alternatives: select in favour of the link that maintains an index balance of nodes in the minimum spanning tree.
3. Since the selected link is of Type 2, return to Step 4 for updating lengths.

Step 7: Stop minimum spanning tree has been obtained.

Let us apply the above approach to the reconstructed network in Figure 5.13. Initially, all links are non-basic, hence values of all $S_{1,k} = S_{1'k} = 0$ for k = 2,3,4. Links lengths arranged in nondecreasing order are:

$$3(1,2), 3(1',2), 4(2,3), 5(1,3), 5(3,4), 5(1'3), \\ 6(1,4), 6(1',4), 6(2,4). \tag{5.22}$$

Note link 3(1,2) will be selected; however, this link is Type 2. Therefore the value of the link (1',2) will change from 3 to ∞. This change of length will alter the order, and the new arrangement of arrangement in nondecreasing order will be as shown in eqn (5.23), given below:

$$3(1,2), 4(2,3), 5(1,3), 5(3,4), 5(1'3), 6(1',4), 6(1,4), 6(2,4), \infty(1',2). \tag{5.23}$$

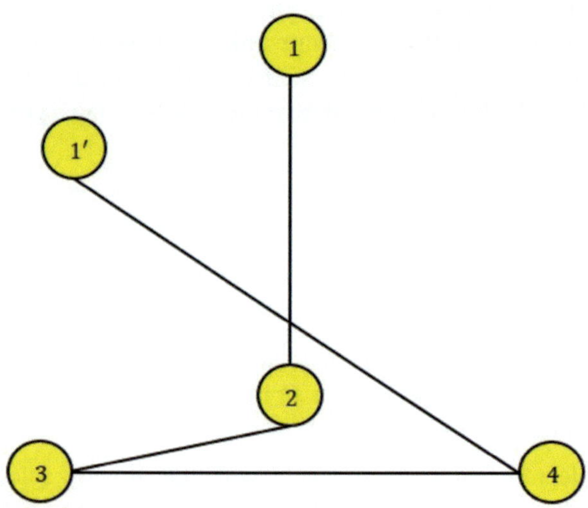

Figure 5.15 The minimum spanning tree of Figure 5.13.

Next link (2,3) of length 4 will be selected, which is of Type 1; hence, we can move to consider the next link of higher length. Next is the link (1,3), which will be discarded as it makes a cycle and when we move to the next one, which is (3,4). Therefore, the selected links are 3(1,2), 4(2,3), 5(3,4). The next link is 5(1',3) can be selected; however, it will increase the index to 3 of the node 3. Therefore, readjustment by the index balancing theorem will be required and the link selected will be 6(1',4). The final MST will be as shown in Figure 5.15.

When nodes 1 and 1' merge, Figure 5.15 will become as shown in Figure 5.16.

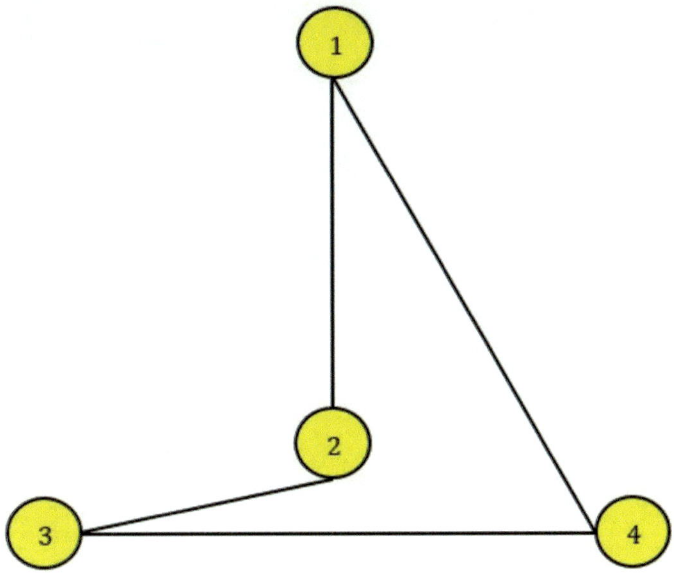

Figure 5.16 The required travelling salesman tour.

5.4.3 Numerical illustration of the proposed approach

Let us consider *G*(8,14) network (Figure 5.17) that was used by Munapo and Kumar (2022) (page 163) to develop a binary variable model that was used to illustrate the transformation discussed in Section 5.4.2. Here, we find the minimum travelling tour using the method discussed in this section and establish that the index restricted minimum spanning tree is equivalent to the minimum salesman tour.

Since node 8 has the minimum number of edges emanating from it, normally we will select node as the node p in the reconstructed network. However, the travelling salesman tour will pass through all nodes; we can select any other node as the node p. Let us select node 1 to be duplicated as node 1' and it will require three links (1,2), (1,3) and (1,4) to be duplicated in the reconstructed network $G((8+1),(14+3)) = G(9,17)$ network as shown in Figure 5.18.

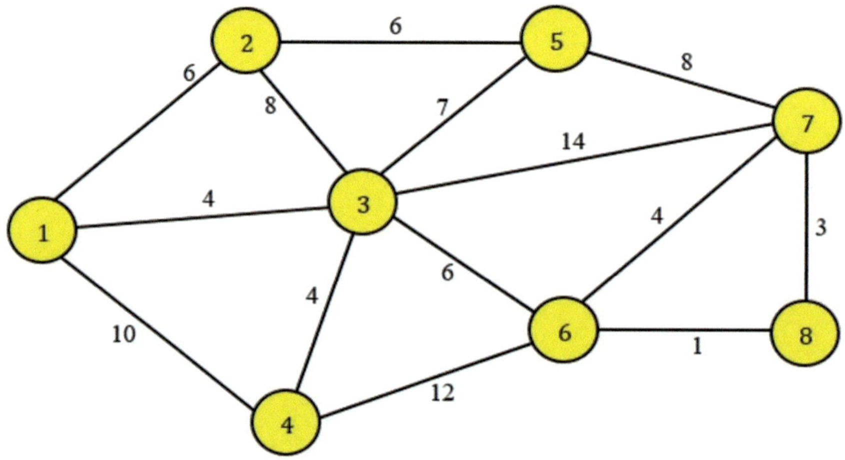

Figure 5.17 *G*(8,14) for the minimum salesman tour.

Table 5.8 Link lengths of links in Figure 5.18.

From/to	1	1'	2	3	4	5	6	7	8
1	-	-	6 or ∞	4 or ∞	10 or ∞	∞	∞	∞	∞
1'	-	-	6 or ∞	4 or ∞	10 or ∞	∞	∞	∞	∞
2	6 or ∞	6 or ∞	-	8	∞	6	∞	∞	∞
3	4 or ∞	4 or ∞	8	-	4	7	6	14	∞
4	10 or ∞	10 or ∞	∞	4	-	∞	12	∞	∞
5	∞	∞	6	7	∞	-	∞	8	∞
6	∞	∞	∞	6	12	∞	-	4	2
7	∞	∞	∞	14	∞	8	4	-	3
8	∞	∞	∞	∞	∞	∞	2	3	-

The reconstructed network *G*(9,17) is shown in Figure 5.17.

Here Type 1 links are: 8(2,3), 6(2,5), 4(3,4), 7(3,5), 6(3,6), 14(3,7), 12(4,6), 8(5,7), 4(6,7), 2(6,8) and 3(7,8).

Type 2 links are: 6 or ∞ {(1,2), or $(1',2)$}, 4 or ∞ {(1,3) or $(1',3)$}, 10 or ∞ {(1,4) or $(1',4)$}.

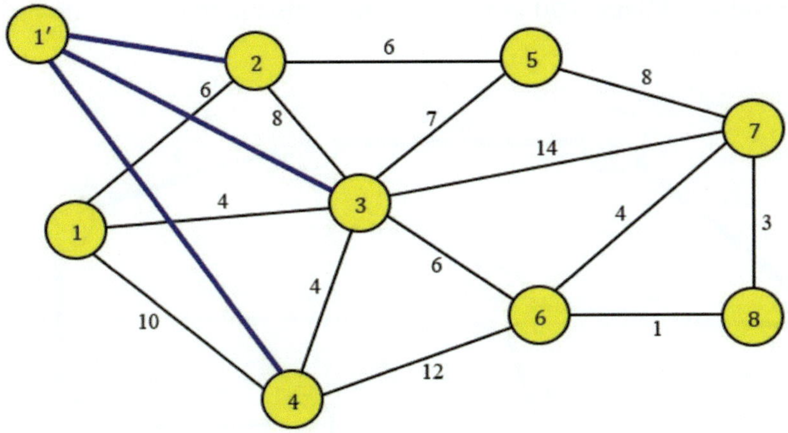

Figure 5.18 Modified network.

Rearranging them in nondecreasing order, we have: Type 1: {2(6,8), 3(7,8), 4(6,7),4(3,4)}, Type 1: 4 or ∞ {(1,3) or (1′,3)}, Type 2: {6(2,5), 6(3,6)}, Type 1:{ 6 or ∞ (1,2), or (1′,2)}, Type 2: {7(3,5), 8(2,3), 8(5,7)}, Type 1: {10 or ∞ {(1,4) or (1′,4)}, Type 2: {12(4,6), 14(3,7)}.

This results in the following minimum spanning tree shown in Figure 5.19, which is also the minimum spanning path, requiring no application of the index balancing theorem. When nodes 1 and 1′ are merged, it becomes the minimum travelling salesman tour.

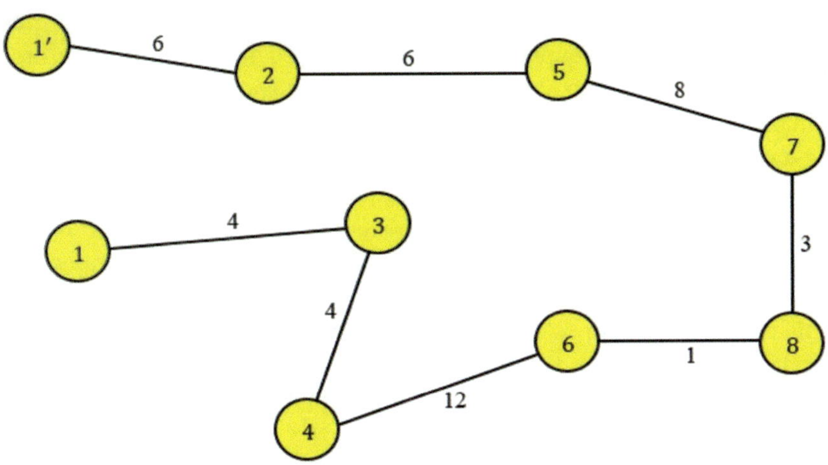

Figure 5.19 Minimum spanning tree.

This results in Figure 5.20 as the required minimum travelling salesman tour.

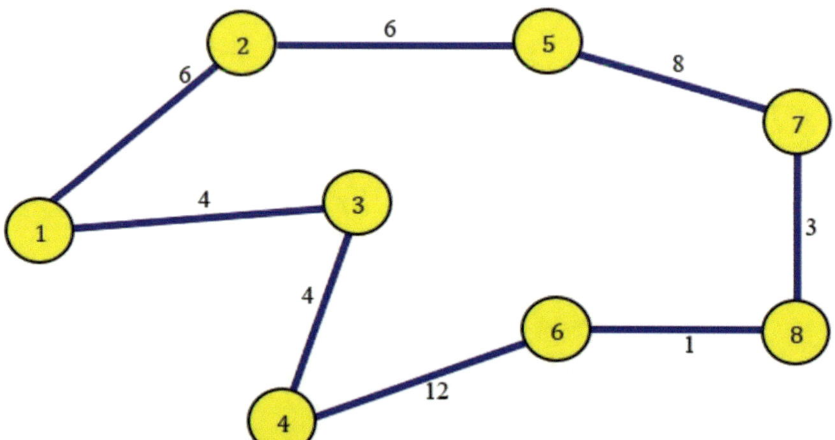

Figure 5.20 The minimum travelling salesman tour.

5.5 Concluding Remarks

In this chapter, for a given connected network, we have presented two recent approaches to find the minimum salesman tour. These approaches were developed by Munapo (2020) and Kumar et al. (2016, 2018, 2019). The title of this chapter raised a question, and our belief is based on the extent that methods discussed in this chapter cannot be classified as NP hard; therefore, we believe that the travelling salesman tour problem should not be classified as a NP hard problem? Of course, further investigations are necessary before reaching a conclusion.

All methods discussed in this chapter require proper computational support to perform some meaningful experiments on larger problems to check performances on these new methods. Studies are also required for comparison of some of the existing methods with these new approaches.

In this chapter, we also established a relationship between the minimum spanning tree, the minimum spanning path and the minimum salesman tour and established an equivalence between the minimum spanning path and the minimum travelling salesman tour.

References

G. Finke, A. Claus, and E. Gunn, A two-commodity network flow approach to the travelling salesman problem, Combinatorics, Graph Theory and Computing, Proc of the 14th South-eastern Conference, Atlantic University, Florida, 1983.

B. Gavish, and S.C. Graves, The travelling salesman problem and related problems, Working Paper OR-078-78, 1978, Operations Research Centre, MIT, Cambridge, MA

S. Kumar, E. Munapo, M. Lesaoana, and P. Nyamgure, A minimum spanning tree-based approach to the travelling salesman tour, Opsearch, Vol 55, No 1, pp150-164, 2018. {DOI 10.1007/s12597-017-0318-5pp1-15.2017}

S. Kumar, E. Munapo, M. Lesaoana, and P. Nyamgure, Is the travelling salesman actually NP Hard?, Chapter 3 in Engineering and Technology: Recent Innovations and Research, Editor Ashok Matani, International Research Publication House, ISBN 978-93-86138-06-4, pp37-58, 2016.

S. Kumar, E. Munapo, C. Siguake, and M. Al-Rabeeah, The minimum spanning tree with node index <=2 is equivalent to the minimum travelling salesman tour, Chapter 8 in Mathematics in Engineering Sciences: Novel theories, Technologies and Applications, ISBN 9781351266321, Editor, Mange Ram, CRC Series (http://www.crcpress.com), pp 227-244, 2019.

S. Kumar and E. Munapo, Innovative ways of developing and using specific purpose alternatives for solving some hard combinatorial network routing and ordered optimization problems, AppliedMath 2024, 4, 791-805. https://doi.org/10.3390/appliedmath4020042.

L.K. Lenstra, and A.G.H. Rinnooykan, Some applications of the travelling salesman problem, Operations Research Quarterly, 26, pp717-33, 1975.

C.E. Miller, A.W. Tucker, and R.A. Zemlin, Integer programming formulation of the travelling salesman problem, JACM 3, pp 326-329, 1960.

E. Munapo, Solving the binary linear programming model in Polynomial time, Americal Journal of OR, 6, 1-7, 2016, http://dx.doi.org/10.4236/ajor.2016.61001.

E. Munapo, and S. Kumar, Linear integer programming: Theory, Applications, Recent Developments, De Gruyter, ISBN 978-3-11-070292-7, 2022.

E. Munapo, Development of a dummy guided formulation and exact solution method for TSP, Eastern-European Journal of Enterprise Technologies, pp12-19, 2020.

E. Munapo, S. Kumar, M. Lesaoana, and P. Nyamgure, A minimum spanning tree with node index <=2, ASOR Bulletin, Vol. 34, Issue 1, pp1-14, 2016. (www.asor.org.au)

G.L. Nemhauser, and L.A. Wolsey, Integer and combinatorial optimization, Wiley Inter-Science Series in Discrete Mathematics and Optimization, North Holland, Amsterdam, 1989.

C.H. Papadimitriou, K. Steiglitz, Combinatorial optimization: Algorithms and Complexity, Dover Publications, 1998.

A. Schrijver, Theory of Linear and Integer Programming, John Wiley and Sons, ISBN 978-0-471-98232-6, 1998.

S. Vajda, Mathematical Programming, Addison-Wesley, London, 1961.

R.T. Wong, Integer programming formulation of the travelling salesman problem, Proc. IEEE Conf. on Circuits and Computers, pp 149-152, 1980.

6

TANYAKUMU Labelling Method: Exact Algorithm for the Travelling Salesman Problem (TSP) and the Equality Generalized Travelling Salesman Problem (E-GTSP)

Abstract

This chapter considers the travelling salesman problem (TSP) and TSP variant namely, the equality generalized travelling salesman problem (E-GTSP). These problems can be modelled using a network hence they are considered to be network optimization problems. Due to the complexity of the problem and its variants, several solution methods have been developed to solve the TSP thus both exact and heuristic solution methods. In this chapter, recently developed exact labelling method is presented to solve the TSP and E-GTSP after minor modifications. The method solves the TSP and E-GTSP in a reasonable number of iterations. The idea behind the proposed method is explained and illustrated through numerical examples.

Keywords: Network reconstruction, link-weight modification, travelling salesman problem, equality generalized travelling salesman problem, optimal tour, exact method.

6.1 Introduction

The travelling salesman problem (TSP) is a network-based combinatorial optimization problem. In this problem, a salesman is supposed to visit n-cities exactly once and return to the starting city called the origin while minimizing the total tour weight. The weight may be distance, time, probability, and cost, among others. TSP has various applications not limited to X-ray crystallography, overhauling turbine engines (Tawanda et al., 2023), DNA sequencing, logistics, and drilling of circuit boards.

6.1.1 Some variants of the travelling salesman problem

The travelling salesman problem gives birth to several variants as technology changes. Some of the variants include, constrained TSP (Tóth, 1988), symmetric TSP (Dell'Amico et al., 1995), asymmetric TSP (Ascheuer, 2001), clustered TSP (Laporte & Palekar, 2002), time-dependent TSP (Arigliano et al., 2018), family TSP (Bernardino & Paias, 2018), multiple TSP (Hossam et al., 2020), quadratic TSP (Woods and Punnen, 2020), maximum scatter TSP (Ahmed, 2020), price collecting TSP (Clímaco et al., 2021), equality generalized TSP (Tawanda et al., 2022), generalized TSP (Asef-Vaziri et al., 2022), flying sidekick TSP (Boccia et al., 2021) and probabilistic TSP (Santini et al., 2022), among others. All these variants prove that the TSP is complex and dynamic in nature.

6.1.2 Some solution methods for the travelling salesman problem

The complexity of the TSP has led to the development of various solution strategies or algorithms such as the following, dynamic programming (Bellman, 1962), genetic algorithms (Larranaga et al., 1999), discrete spider monkey optimization methods (Akhand et al., 2020), Harris Hawk optimization strategy (Gharehchopogh and Abdollahzadeh, 2022), Cuckoo search algorithms (Reda et al., 2022), Harmon search algorithms (Boryczka and Szwarc, 2019), Brach and cut algorithms (Hernández-Pérez and Salazar-González, 2022), branch and bound (Munapo, 2013), artificial bee colony optimization (Fairee et al., 2019), greedy randomized search (Júnior and Guimarães, 2019), nearest neighbour search heuristics (Asani and Okeyinka, 2020), evolutionary algorithms (Michalak, 2021), TANYAKUMU labelling method (Asef-Vaziri et al., 2022), minimum spanning tree-based heuristic (Kumar et al., 2018) and discrete flower pollination algorithms (Santini et al., 2022) among others.

6.2 TANYAKUMU Labelling Method

This is an exact solution method that was developed by Tawanda et al. (2022) to determine the minimum cost tour to the TSP. The method was later extended to solve the equality generalized travelling salesman problem (E-GTSP) and also to solve shortest path problems (SPP) in directed networks. The algorithm is simple and flexible in that it can be easily modified to solve related problems optimally.

6.2.1 Definition of terms

The following definitions are associated with the development of the algorithm.

Definition 1: Sets of nodes

L_0 = The set of intermediate nodes given by $L_0 = \{2, 3, \ldots, k\}$. This set turns to an empty, thus $L_0 = \{\emptyset\}$ when all the nodes are visited. The nodes in this set must be included exactly once in each and every tour. $L^*{}_0$ = The set that contains the source and destination nodes of the travelling salesman problem, thus $L^*{}_0 = \{1\}$. The nodes of this set should be at the extremes of each and every valid tour.

Definition 2: Node label

Given a network with node 1 as the origin city and other intermediate nodes as $\{2, 3, 4, \ldots, n\}$. A generalized node label is given by $w_{(1,j)^z}$, where j is the currently added node to the partial tour $(1, j)$ with z intermediate nodes. W is the total weight or cost of travelling from node 1 to the currently added node j through z intermediate nodes. Another label (∞) is used to restrict adding node j to the node label if node j is already part of the node label.

Definition 3: Minimum weight label

For every node j the minimum weight label is given by $w^*{}_{(1,k)^z} = \text{Min}\left\{w_{(1,j)^z}\right\}$. These minimum weight labels are used as seed for the next iteration, thus they are selected as fit labels for modification.

Definition 4: Minimum weight label modification

Node labels are modified as the algorithm iterates, thus as the algorithms adds new possible partial tour neighbour nodes. A minimum weight label $w^*{}_{(1,k)^z}$ can be modified to $w_{(1,j)^{z+1}}$; thus, $w_{(1,j)^{z+1}} = w^*{}_{(1,k)^z} + c_{\{k,j\}}$, where c is the weight of the edge $\{k, j\}$, where k and j are the previously and currently visited nodes respectively, $k, j \in L_0$.

6.2.2 The algorithm

In this section, the TANYAKUMU labelling algorithm to solve the original travelling salesman problem is presented. The steps of the algorithm are presented below. Numerical examples are also presented to explain how the method computes the optimal solution.

Algorithm 6. 1. TANYAKUMU Labelling Method for the TSP

Initialization. For any given network represent home city node by $\mathbf{1}$. The set of all unvisited nodes is given by $L_O = \{2,\ 3, \ldots, n\}$. To initialize the algorithm label all nodes with $w_{(1,j)^z}$ for all $j \in L_O$ that are directly connected to the home city (node $\mathbf{1}$). Where w is the weight of the shortest path $(1, j)$ with intermediate nodes z, **initially** $z = 0$. At the initialization stage $w^*{}_{(1,j)^0} = w_{(1,j)^0}$

Step 1. If $z < (n - 2)$, Assign $j \to k$, **where** k is the previous added node thus changing the label $w^*{}_{(1,j)^z}$ to $w^*{}_{(1,k)^z}$. For all $j \in L_O$ that is part of the minimum weight label $w^*{}_{(1,k)^z}$ label j with (∞). Else, modify $w^*{}_{(1,k)^z}$ to $w_{(1,j)^{z+1}}$ and label j with $w_{(1,j)^{z+1}}$ where $w_{(1,j)^{z+1}} = w^*{}_{(1,k)^z} + c_{\{k,j\}}$.

Step 2. For all $j \in L_O$ determine the minimum weight label using the following formula: $w^*{}_{(1,j)^{z+1}} = \mathrm{Min}\left\{ w_{(1,j)^{z+1}} \right\}$ and if node j has ties consider all of them.

Step 3. If $z = (n - 2)$ then go to Step 4. Else go to Step 1. When $z = (n - 2)$ it follows that all the labels have all elements of $j \in L_O$ as visited nodes hence $L_O = \{\emptyset\}$.

Step 4. Since $L_1 = \{\emptyset\}$ there no remaining node to visit as a result we are now forced to return back to the home city (node 1). Consider $j \in L^*{}_O = \{1\}$ and go to Steps 1 and 2. Then STOP.

Step 5. The tour corresponding to the minimum weight label $w^*{}_{(1,j)^{n-2}}$ is the minimum weight tour, where $j \in L^*{}_O = \{1\}$.

Example 6.1

Consider the following 4 nodes travelling salesman problem given in Figure 6.1. Determine the minimum cost tour that visits all the nodes exactly once and return to the origin node.

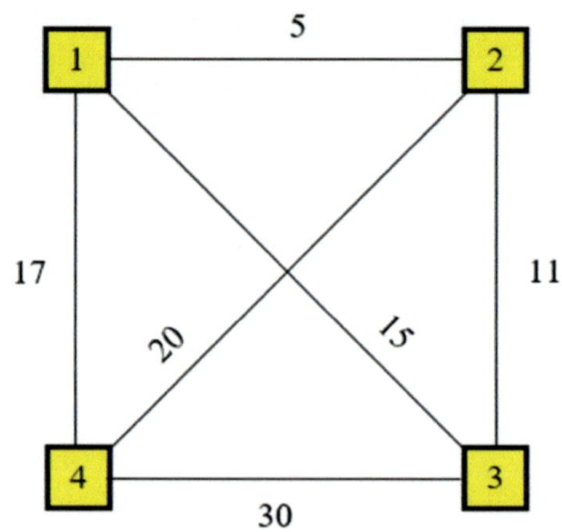

Figure 6.1 Four nodes of travelling salesman problem.

Solution

Initialization: To initialize the search, connect node 1, the origin to all other nodes in the network thus determining ${w^*}_{(1,j)^0}$ node labels, where j is the neighbour node for the origin node. Set $j \to k$ and go to iteration 1.

Iteration 1: Node labels ${w^*}_{(1,k)^0}$ are modified to get new node labels $w_{(1,j)^1}$ by connecting k to all j not part of the particular node label, else the new node label is (∞). Next, compute ${w^*}_{(1,j)^1}$ the minimum weight partial tours for all the nodes except the origin node as shown in Table 6.1. ${w^*}_{(1,j)^1}$ is used as a seed for the next iteration. Set $j \to k$ and go to iteration 2.

Table 6.1 Partial tours after iteration 1.

${w^*}_{(1,k)^0}$	j=1	j=2	j=3	j=4
$5_{(1,2)^0}$	∞	∞	$25_{(1,2,3)^1}$	$16_{(1,2,4)^1}$
$17_{(1,3)^0}$	∞	$37_{(1,3,2)^1}$	∞	$47_{(1,3,4)^1}$
$15_{(1,4)^0}$	∞	$26_{(1,4,2)^1}$	$45_{(1,4,3)^1}$	∞
${w^*}_{(1,j)^1}$	–	$26_{(1,4,2)^1}$	$25_{(1,2,3)^1}$	$16_{(1,2,4)^1}$

Iteration 2: Node labels ${w^*}_{(1,k)^1}$ are modified to get new node labels $w_{(1,j)^2}$ by connecting k to all j not part of the particular node label, else the new node label is (∞). Then compute ${w^*}_{(1,j)^2}$ the minimum weight partial tours for all the nodes except the origin node as shown in Table 6.2. ${w^*}_{(1,j)^2}$ is used as a seed for the next iteration. Set $j \to k$ and go to iteration 3.

Table 6.2 Partial tours after iteration 2.

${w^*}_{(1,k)^1}$	j=1	j=2	j=3	j=4
$26_{(1,4,2)^1}$	∞	∞	$46_{(1,4,2,3)^2}$	∞
$25_{(1,2,3)^1}$	∞	∞	∞	$55_{(1,2,3,4)^2}$
$16_{(1,2,4)^1}$	∞	∞	$46_{(1,2,4,3)^2}$	∞
${w^*}_{(1,j)^2}$	–	–	All Routes	$55_{(1,2,3,4)^2}$

Iteration 3: Since $j \in L_0 = \{\emptyset\}$, all the nodes in the network have been visited exactly once. The salesman has to return to the origin node, thus modifying node labels ${w^*}_{(1,k)^2}$ by connecting them back to the origin node to obtain a complete tour as shown in Table 6.3. Compute ${w^*}_{(1,j)^3}$ for the origin node. ${w^*}_{(1,j)^3}$ is the optimal solution to the travelling salesman problem.

Minimum cost tours are 1→4→2→3→1 and 1→2→4→3→1 with tour distance of 54 units. Figure 6.2 shows the minimum tours on the TSP network.

Table 6.3 Partial tours after iteration 3.

$w^*_{(1,k)^2}$	j=1	j=2	j=3	j=4
$\mathbf{46}_{(1,4,2,3)^2}$	$63_{(1,4,2,3,1)^3}$	∞	∞	∞
$\mathbf{46}_{(1,2,4,3)^2}$	$63_{(1,2,4,3,1)^3}$	∞	∞	∞
$\mathbf{55}_{(1,2,3,4)^2}$	$70_{(1,2,3,4,1)^3}$	∞	∞	∞
$w^*_{(1,j)^3}$	Tie optimal tours	–	–	–

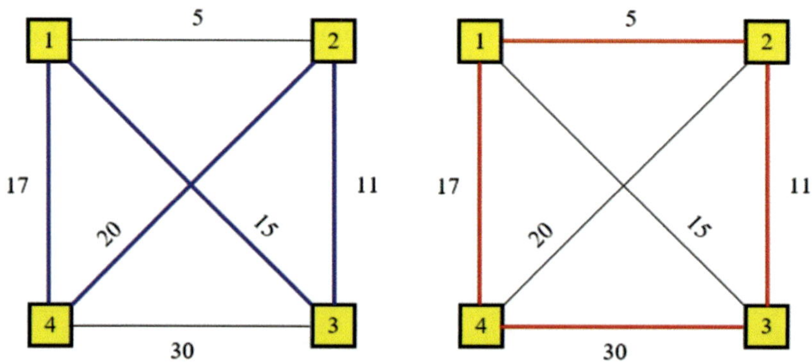

Figure 6.2 Solution to four nodes travelling salesman problem.

Example 6.2

Given in Figure 6.3 is a travelling salesman problem. The network shows how the nodes are connected as well as distances between them. The problem is to

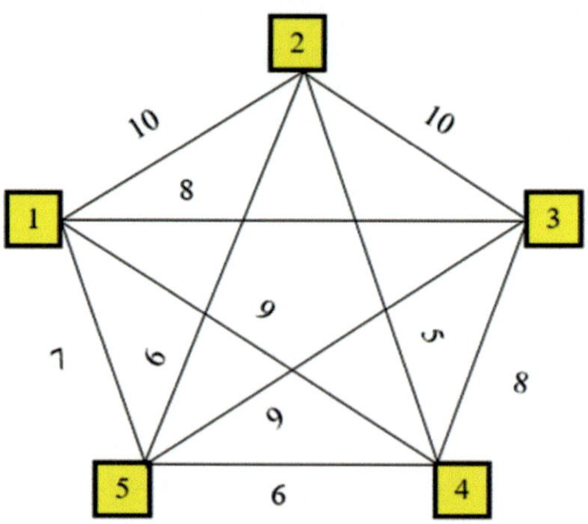

Figure 6.3 Five nodes travelling salesman problem.

determine a cycle that starts from any node and visit all the remaining nodes exactly once and then return to the origin node while minimizing travelled distance.

Solution

Initialization: To initialize the search, node 1, the origin is linked to all other nodes in the network, thus determining $w^*{}_{(1,j)^0}$ node labels, where j is the neighbour node for the origin. Set $j \to k$ and go to iteration 1.

Iteration 1: Node labels $w^*{}_{(1,k)^0}$ are modified to get new node labels $w_{(1,j)^1}$ by linking k to all j not part of the particular node label, else the new node label is (∞). Then compute $w^*{}_{(1,j)^1}$ the minimum weight partial tours for all the nodes except the origin node as shown in Table 6.4. $w^*{}_{(1,j)^1}$ is used as a seed for the next iteration. Set $j \to k$ and go to iteration 2.

Table 6.4 Partial tours after iteration 1.

$w^*{}_{(1,k)^0}$	j=1	j=2	j=3	j=4	j=5
$10_{(1,2)^0}$	∞	∞	$20_{(1,2,3)^1}$	$15_{(1,2,4)^1}$	$16_{(1,2,5)^1}$
$8_{(1,3)^0}$	∞	$18_{(1,3,2)^1}$	∞	$16_{(1,3,4)^1}$	$17_{(1,3,5)^1}$
$9_{(1,4)^0}$	∞	$14_{(1,4,2)^1}$	$17_{(1,4,3)^1}$	∞	$15_{(1,4,5)^1}$
$7_{(1,5)^0}$	∞	$13_{(1,5,2)^1}$	$16_{(1,5,3)^1}$	$13_{(1,5,4)^1}$	∞
$w^*{}_{(1,j)^1}$	–	$13_{(1,5,2)^1}$	$16_{(1,5,3)^1}$	$13_{(1,5,4)^1}$	$15_{(1,4,5)^1}$

Iteration 2: Node labels $w^*{}_{(1,k)^1}$ are modified to get new node labels $w_{(1,j)^2}$ by connecting k to all j not part of the particular node label, else the new node label is (∞). Then compute $w^*{}_{(1,j)^2}$ the minimum weight partial tours for all the nodes except the origin node as shown in Table 6.5. $w^*{}_{(1,j)^2}$ is used as a seed for the next iteration. Set $j \to k$ and go to iteration 3.

Table 6.5 Partial tours after iteration 2.

$w^*{}_{(1,k)^1}$	j=1	j=2	j=3	j=4	j=5
$13_{(1,5,2)^1}$	∞	∞	$23_{(1,5,2,3)^2}$	$18_{(1,5,2,4)^2}$	∞
$16_{(1,5,3)^1}$	∞	$26_{(1,5,3,2)^2}$	∞	$24_{(1,5,3,4)^2}$	∞
$13_{(1,5,4)^1}$	∞	$18_{(1,5,4,2)^2}$	$21_{(1,5,4,3)^2}$	∞	∞
$15_{(1,4,5)^1}$	∞	$21_{(1,4,5,2)^2}$	$24_{(1,4,5,3)^2}$	∞	∞
$w^*{}_{(1,j)^2}$	–	$18_{(1,5,4,2)^2}$	$21_{(1,5,4,3)^2}$	$18_{(1,5,2,4)^2}$	–

Iteration 3: Node labels $w^*{}_{(1,k)^2}$ are modified to get new node labels $w_{(1,j)^3}$ by connecting k to all j not part of the particular node label, else the new

Table 6.6 Partial tours after iteration 3.

$w^*_{(1,k)^2}$	j=1	j=2	j=3	j=4	j=5
$\mathbf{18}_{(1,5,4,2)^2}$	∞	∞	$28_{(1,5,4,2,3)^3}$	∞	∞
$\mathbf{21}_{(1,5,4,3)^2}$	∞	$31_{(1,5,4,3,2)^3}$	∞	∞	∞
$\mathbf{18}_{(1,5,2,4)^2}$	∞	∞	$26_{(1,5,2,4,3)^3}$	∞	∞
$w^*_{(1,j)^3}$	–	$31_{(1,5,4,3,2)^3}$	$26_{(1,5,2,4,3)^3}$	–	–

node label is (∞). Then compute $w^*_{(1,j)^3}$ the minimum weight partial tours for all the nodes except the origin node. $w^*_{(1,j)^3}$ is used as a seed for the next iteration as shown in Table 6.6. Set $j \to k$ and go to iteration 4.

Iteration 4: Since $j \in L_0 = \{\emptyset\}$, thus all nodes in the network have been visited exactly once. The salesman has to return to the origin node, thus modifying node labels $w^*_{(1,k)^3}$ by connecting them back to the origin node to obtain a complete tour as shown in Table 6.7. Compute $w^*_{(1,j)^4}$ for the origin node. $w^*_{(1,j)^4}$ is the optimal solution to the travelling salesman problem.

Table 6.7 Partial tours after iteration 4.

$w^*_{(1,k)^3}$	j=1	j=2	j=3	j=4	j=5
$\mathbf{31}_{(1,5,4,3,2)^3}$	$41_{(1,5,4,3,2,1)^4}$	∞	∞	∞	∞
$\mathbf{26}_{(1,5,2,4,3)^3}$	$34_{(1,5,2,4,3,1)^4}$	∞	∞	∞	∞
$w^*_{(1,j)^4}$	$34_{(1,5,2,4,3,1)^4}$	–	–	–	–

Minimum cost tours are $1 \to 5 \to 2 \to 4 \to 3 \to 1$ with a total distance of 34 units. Figure 6.4 shows the minimum tours on the TSP network.

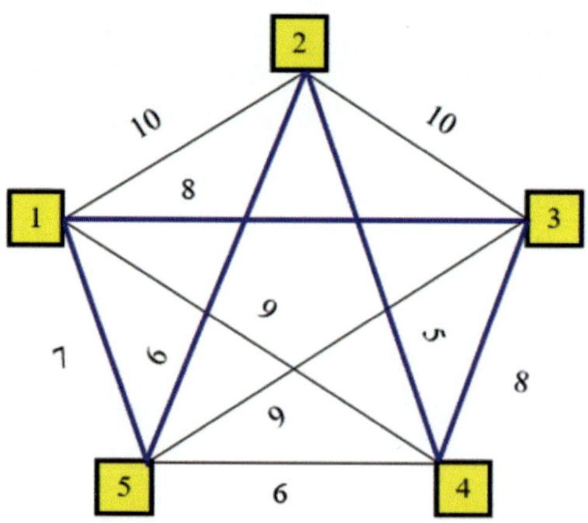

Figure 6.4 Five nodes travelling salesman problem.

6.3 Equality Generalized Travelling Salesman Problem (E-GTSP)

In this variant of TSP, called the E-GTSP, the interconnected nodes of the TSP are grouped to form clusters. The salesman is required start from the origin node and visit exactly one node in each and every cluster and return to the origin node while minimizing total travelled distance. Mathematically, E-GTSP can be defined as $G = (V, E)$ where V is the set of all vertices in the network and E is the set of all edges in the network. Set V is partitioned into m clusters given by $V_1,\ V_2, \ldots, V_m$ in such a way that $V = V_1 \cup V_2, \ldots, \cup V_m$ where $V_i \cap V_j = \emptyset$ for all $i \neq j$.

6.3.1 Modified TANYAKUMU labelling algorithm for E-GTSP

In this section, the labelling method is modified to solve the E-GTSP. The steps of the modified algorithm are presented below.

Algorithm 6.2. Modified TANYAKUMU labelling method for E-GTSP

Initialization Label all vertices with label $w_{(1,j)^z}$ for all $j \in \mathbf{V{=}V_1 \cup V_2 \ldots \cup V_m}$. At the initialization step $w^*{}_{(1,j)^0} = w_{(1,j)^0}$.

Step 1. For all $j \in w^*{}_{(1,j)^z}$ and V_i where $i = 1,\ 2,\ 3, \ldots, m$

a) Label V_i with label (∞)

b) Else label V_i with label $w_{(1,j)^{z+1}}$ where $w_{(1,j)^{z+1}} = w^*{}_{(1,k)^z} + c_{(k,j)}$.

Step 2. For all $j \in V_i$ where $i = 1,\ 2,\ 3, \ldots, m$. Determine the minimum weight label using the following formula: $w^*{}_{(1,j)^{z+1}} = Min \left\{ w_{(1,j)^{z+1}} \right\}$

Step 3. If $z < m - 1$. Assign $j \to k$ and go to step 1, Else $z = m - 1 \Rightarrow$ all V_i are visited exactly once. Go to step 4.

Step 4. Return to $j = 1$(Home city) and go to step 2, then STOP. The tour corresponding to the minimum weight label $w^*{}_{(1,j)^{z=m}}$ is the minimum weight tour.

Example 6.3

Consider the following E-GTSP given in Figure 6.5 with eight vertices that are partitioned to form two clusters thus $V_1 = \{1,\ 2, 3, 4, 5\}$ and $V_2 = \{6,\ 7,\ 8,\ 9\}$. Node 0 is the home city. Table 6.8 below shows the clusters and the distances between all the vertices in the problem.

Solution

Initialization: To initialize the search, node 0 the origin is connected to all other nodes in all the clusters thus determining $w^*{}_{(1,j)^0}$ node labels, where j are the neighbour nodes for the origin node. In this problem, $m = 2$ and

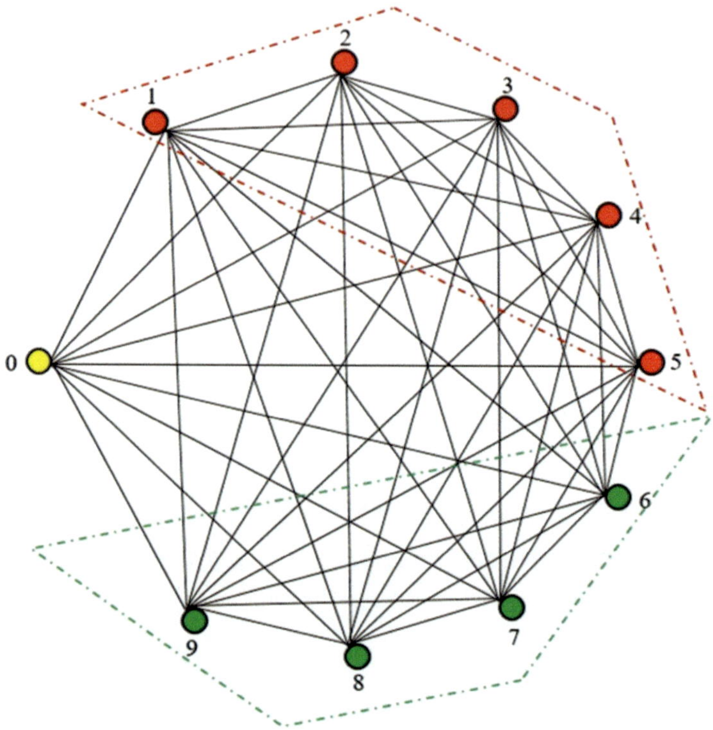

Figure 6.5 Two clusters E-GTSP.

Table 6.8 Distance matrix for the two clusters E-GTSP problem.

	Origin	Cluster 1					Cluster 2			
	$j=0$	$j=1$	$j=2$	$j=3$	$j=4$	$j=5$	$j=6$	$j=7$	$j=8$	$j=9$
	0	**1**	**2**	**3**	**4**	**5**	**6**	**7**	**8**	**9**
0	∞	12	3	23	1	5	32	56	91	84
1	12	∞	9	18	3	41	45	5	79	13
2	3	9	∞	89	56	21	12	49	70	11
3	23	18	89	∞	87	46	75	17	63	50
4	1	3	56	87	∞	55	22	86	62	44
5	5	41	21	46	55	∞	21	76	84	27
6	32	45	12	75	22	21	∞	11	54	60
7	56	5	49	17	86	76	11	∞	41	77
8	91	79	70	63	62	84	54	41	∞	10
9	84	13	11	50	44	27	60	77	10	∞

$z=0$ intermediate nodes on all partial tours. While $(z \leq m-1)$ assign $j \to k$ and go to iteration 1.

Table 6.9 Partial tours after iteration 1.

	Cluster 1					Cluster 2			
$w^*{}_{(1,k)^0}$	$j=1$	$j=2$	$j=3$	$j=4$	$j=5$	$j=6$	$j=7$	$j=8$	$j=9$
$\mathbf{12}_{(0,1)^0}$	∞	∞	∞	∞	∞	$57_{(0,1,6)^1}$	$17_{(0,1,7)^1}$	$91_{(0,1,8)^1}$	$25_{(0,1,9)^1}$
$\mathbf{3}_{(0,2)^0}$	∞	∞	∞	∞	∞	$15_{(0,2,6)^1}$	$52_{(0,2,7)^1}$	$73_{(0,2,8)^1}$	$14_{(0,2,9)^1}$
$\mathbf{23}_{(0,3)^0}$	∞	∞	∞	∞	∞	$98_{(0,3,6)^1}$	$40_{(0,3,7)^1}$	$86_{(0,3,8)^1}$	$73_{(0,3,9)^1}$
$\mathbf{1}_{(0,4)^0}$	∞	∞	∞	∞	∞	$23_{(0,4,6)^1}$	$87_{(0,4,7)^1}$	$63_{(0,4,8)^1}$	$45_{(0,4,9)^1}$
$\mathbf{5}_{(0,5)^0}$	∞	∞	∞	∞	∞	$26_{(0,5,6)^1}$	$81_{(0,5,7)^1}$	$89_{(0,5,8)^1}$	$32_{(0,5,9)^1}$
$\mathbf{32}_{(0,6)^0}$	$77_{(0,6,1)^1}$	$44_{(0,6,2)^1}$	$107_{(0,6,3)^1}$	$54_{(0,6,4)^1}$	$53_{(0,6,5)^1}$	∞	∞	∞	∞
$\mathbf{56}_{(0,7)^0}$	$61_{(0,7,1)^1}$	$105_{(0,7,2)^1}$	$73_{(0,7,3)^1}$	$142_{(0,7,4)^1}$	$132_{(0,7,5)^1}$	∞	∞	∞	∞
$\mathbf{91}_{(0,8)^0}$	$170_{(0,8,1)^1}$	$161_{(0,8,2)^1}$	$154_{(0,8,3)^1}$	$153_{(0,8,4)^1}$	$175_{(0,8,5)^1}$	∞	∞	∞	∞
$\mathbf{84}_{(0,9)^0}$	$97_{(0,9,1)^1}$	$95_{(0,9,2)^1}$	$134_{(0,9,3)^1}$	$128_{(0,9,4)^1}$	$111_{(0,9,5)^1}$	∞	∞	∞	∞
$w^*{}_{(1,j)^1}$	$61_{(0,7,1)^1}$	$44_{(0,6,2)^1}$	$73_{(0,7,3)^1}$	$54_{(0,6,4)^1}$	$53_{(0,6,5)^1}$	$15_{(0,2,6)^1}$	$17_{(0,1,7)^1}$	$63_{(0,4,8)^1}$	$14_{(0,2,9)^1}$

Iteration 1: Node labels $w^*{}_{(1,k)^0}$ are modified to get new node labels $w_{(1,j)^1}$ by connecting k to all j not part of the particular node label, else the new node label is (∞). Then compute $w^*{}_{(1,j)^1}$ the minimum weight partial tours for all the nodes except the origin node as shown in Table 6.9. The number of intermediate nodes is given by $z = 1$ thus $(z \leq m - 1)$ implying that the partial tours have not yet visited all the clusters. $w^*{}_{(1,j)^1}$ is used as a seed for the next iteration. Set $j \rightarrow k$ and go to iteration 2.

Iteration 2: Exactly one node in every cluster has been visited. The salesman has to return to the origin node, thus modifying node labels $w^*{}_{(1,k)^1}$ by connecting them back to the origin node to obtain a complete tour as shown in Table 6.10. We then compute $w^*{}_{(1,j)^2}$ for the origin node. $w^*{}_{(1,j)^2}$ is the optimal solution to the equality generalized travelling salesman problem.

Table 6.10 Partial tours after iteration 2.

	Origin	Cluster 1					Cluster 2			
$w^*{}_{(1,k)^1}$	$j=0$	$j=1$	$j=2$	$j=3$	$j=4$	$j=5$	$j=6$	$j=7$	$j=8$	$j=9$
$61_{(0,7,1)^1}$	$73_{(0,7,1,0)^2}$	∞	∞	∞	∞	∞	∞	∞	∞	∞
$44_{(0,6,2)^1}$	$47_{(0,6,2,0)^2}$	∞	∞	∞	∞	∞	∞	∞	∞	∞
$73_{(0,7,3)^1}$	$96_{(0,7,3,0)^2}$	∞	∞	∞	∞	∞	∞	∞	∞	∞
$54_{(0,6,4)^1}$	$55_{(0,6,4,0)^2}$	∞	∞	∞	∞	∞	∞	∞	∞	∞
$53_{(0,6,5)^1}$	$58_{(0,6,5,0)^2}$	∞	∞	∞	∞	∞	∞	∞	∞	∞
$15_{(0,2,6)^1}$	$47_{(0,2,6,0)^2}$	∞	∞	∞	∞	∞	∞	∞	∞	∞
$17_{(0,1,7)^1}$	$73_{(0,1,7,0)^2}$	∞	∞	∞	∞	∞	∞	∞	∞	∞
$63_{(0,4,8)^1}$	$154_{(0,4,8,0)^2}$	∞	∞	∞	∞	∞	∞	∞	∞	∞
$14_{(0,2,9)^1}$	$98_{(0,2,9,0)^2}$	∞	∞	∞	∞	∞	∞	∞	∞	∞
$w^*{}_{(1,j)^2}$	$47_{(0,2,6,0)^2}$	-	-	-	-	-	-	-	-	-

Minimum cost tour is given by 0→2→6→0 with a total distance of 47 units. Figure 6.6 shows the minimum tours on the E-GTSP network.

Example 6.4: The nodes of the E-GTSP problem in example 3 are regrouped to form 3 clusters $V_1 = \{1, 2, 3\}$, $V_2 = \{4, 5, 6\}$ and $V_3 = \{7, 8, 9\}$ as shown in Table 6.11. The corresponding network showing interconnection between nodes is shown in Figure 6.7.

Solution

Initialization: To initialize the search, node 0, the origin is connected to all other nodes in all the clusters thus determining $w^*{}_{(1,j)^0}$ node labels, where j are the neighbour nodes for the origin nodes. In this problem, $m = 3$ clusters

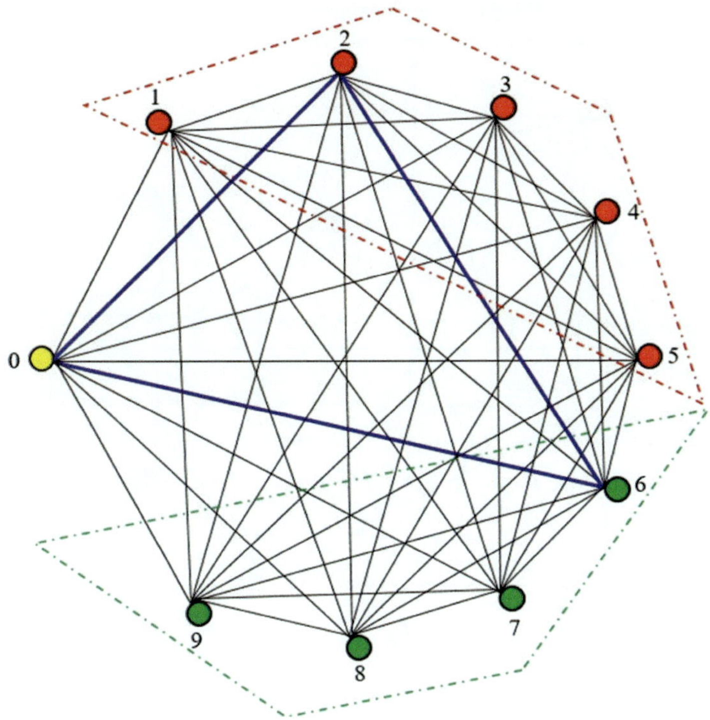

Figure 6.6 Solution to two clusters E-GTSP.

Table 6.11 Distance matrix for the three clusters E-GTSP problem.

	Origin	Cluster 1			Cluster 2			Cluster 3		
	$j=0$	$j=1$	$j=2$	$j=3$	$j=4$	$j=5$	$j=6$	$j=7$	$j=8$	$j=9$
	0	**1**	**2**	**3**	**4**	**5**	**6**	**7**	**8**	**9**
0	∞	12	3	23	1	5	32	56	91	84
1	12	∞	9	18	3	41	45	5	79	13
2	3	9	∞	89	56	21	12	49	70	11
3	23	18	89	∞	87	46	75	17	63	50
4	1	3	56	87	∞	55	22	86	62	44
5	5	41	21	46	55	∞	21	76	84	27
6	32	45	12	75	22	21	∞	11	54	60
7	56	5	49	17	86	76	11	∞	41	77
8	91	79	70	63	62	84	54	41	∞	10
9	84	13	11	50	44	27	60	77	10	∞

and $z = 0$ intermediate nodes on all partial tours. While $(z \leq m - 1)$ assign $j \to k$ and go to iteration 1.

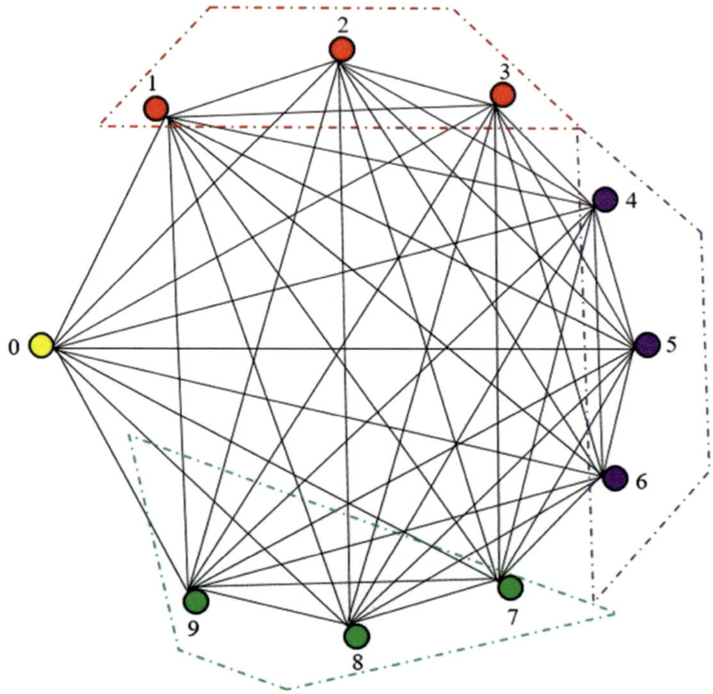

Figure 6.7 Three clusters E-GTSP.

Iteration 1: Node labels ${w^*}_{(1,k)^0}$ are modified to get new node labels $w_{(1,j)^1}$ by connecting k to all j not part of the particular node label, else the new node label is (∞). Then compute ${w^*}_{(1,j)^1}$ the minimum weight partial tours for all the nodes except the origin node as shown in Table 6.12. The number of intermediate nodes is given by $z = 1$ thus $(z \leq m - 1)$ implying that the partial tours have not yet visited all the clusters. ${w^*}_{(1,j)^1}$ is used as a seed for the next iteration. Set $j \rightarrow k$ and go to iteration 2.

Iteration 2: Node labels ${w^*}_{(1,k)^1}$ are modified to get new node labels $w_{(1,j)^2}$ by connecting k to all j not part of the particular node label, else the new node label is (∞). Then compute ${w^*}_{(1,j)^2}$ the minimum weight partial tours for all the nodes except the origin node as shown in Table 6.13. The number of intermediate nodes is given by $z = 2$ thus $(z \leq m - 1)$ implying that the partial tours have not yet visited all the clusters. ${w^*}_{(1,j)^2}$ is used as a seed for the next iteration. Set $j \rightarrow k$ and go to iteration 3.

Table 6.12 Partial tours after iteration 1.

	Cluster 1			Cluster 2			Cluster 3		
$w^*{}_{(1,k)}{}^0$	$j=1$	$j=2$	$j=3$	$j=4$	$j=5$	$j=6$	$j=7$	$j=8$	$j=9$
$\mathbf{12}_{(0,1)}{}^0$	∞	∞	∞	$15_{(0,1,4)}{}^1$	$53_{(0,1,5)}{}^1$	$57_{(0,1,6)}{}^1$	$17_{(0,1,7)}{}^1$	$91_{(0,1,8)}{}^1$	$25_{(0,1,9)}{}^1$
$\mathbf{3}_{(0,2)}{}^0$	∞	∞	∞	$59_{(0,2,4)}{}^1$	$24_{(0,2,5)}{}^1$	$15_{(0,2,6)}{}^1$	$52_{(0,2,7)}{}^1$	$73_{(0,2,8)}{}^1$	$14_{(0,2,9)}{}^1$
$\mathbf{23}_{(0,3)}{}^0$	∞	∞	∞	$110_{(0,3,4)}{}^1$	$69_{(0,3,5)}{}^1$	$98_{(0,3,6)}{}^1$	$40_{(0,3,7)}{}^1$	$86_{(0,3,8)}{}^1$	$73_{(0,3,9)}{}^1$
$\mathbf{1}_{(0,4)}{}^0$	$4_{(0,4,1)}{}^1$	$57_{(0,4,2)}{}^1$	$88_{(0,4,3)}{}^1$	∞	∞	∞	$87_{(0,4,7)}{}^1$	$63_{(0,4,8)}{}^1$	$45_{(0,4,9)}{}^1$
$\mathbf{5}_{(0,5)}{}^0$	$46_{(0,5,1)}{}^1$	$26_{(0,5,2)}{}^1$	$51_{(0,5,3)}{}^1$	∞	∞	∞	$81_{(0,5,7)}{}^1$	$89_{(0,5,8)}{}^1$	$32_{(0,5,9)}{}^1$
$\mathbf{32}_{(0,6)}{}^0$	$77_{(0,6,1)}{}^1$	$44_{(0,6,2)}{}^1$	$107_{(0,6,3)}{}^1$	∞	∞	∞	$43_{(0,6,7)}{}^1$	$86_{(0,6,8)}{}^1$	$92_{(0,6,9)}{}^1$
$\mathbf{56}_{(0,7)}{}^0$	$61_{(0,7,1)}{}^1$	$105_{(0,7,2)}{}^1$	$73_{(0,7,3)}{}^1$	$142_{(0,7,4)}{}^1$	$132_{(0,7,5)}{}^1$	$67_{(0,7,6)}{}^1$	∞	∞	∞
$\mathbf{91}_{(0,8)}{}^0$	$170_{(0,8,1)}{}^1$	$161_{(0,8,2)}{}^1$	$154_{(0,8,3)}{}^1$	$153_{(0,8,4)}{}^1$	$175_{(0,8,5)}{}^1$	$145_{(0,8,6)}{}^1$	∞	∞	∞
$\mathbf{84}_{(0,9)}{}^0$	$97_{(0,9,1)}{}^1$	$95_{(0,9,2)}{}^1$	$134_{(0,9,3)}{}^1$	$128_{(0,9,4)}{}^1$	$111_{(0,9,5)}{}^1$	$144_{(0,9)}{}^1$	∞	∞	∞
$w^*{}_{(1,j)}{}^1$	$4_{(0,4,1)}{}^1$	$26_{(0,5,2)}{}^1$	$51_{(0,5,3)}{}^1$	$15_{(0,1,4)}{}^1$	$24_{(0,2,5)}{}^1$	$15_{(0,2,6)}{}^1$	$17_{(0,1,7)}{}^1$	$63_{(0,4,8)}{}^1$	$14_{(0,2,9)}{}^1$

Table 6.13 Partial tours after iteration 2.

	Cluster 1			Cluster 2			Cluster 3		
$\boldsymbol{w^*}_{(1,k)^1}$	$j=1$	$j=2$	$j=3$	$j=4$	$j=5$	$j=6$	$j=7$	$j=8$	$j=9$
$\mathbf{4}_{(0,4,1)^1}$	∞	∞	∞	∞	∞	∞	$9_{(0,4,1,7)^2}$	$83_{(0,4,1,8)^2}$	$17_{(0,4,1,9)^2}$
$\mathbf{26}_{(0,5,2)^1}$	∞	∞	∞	∞	∞	∞	$75_{(0,5,2,7)^2}$	$96_{(0,5,2,8)^2}$	$37_{(0,5,2,9)^2}$
$\mathbf{51}_{(0,5,3)^1}$	∞	∞	∞	∞	∞	∞	$68_{(0,5,3,7)^2}$	$114_{(0,5,3,8)^2}$	$101_{(0,5,3,9)^2}$
$\mathbf{15}_{(0,1,4)^1}$	∞	∞	∞	∞	∞	∞	$101_{(0,1,4,7)^2}$	$77_{(0,1,4,8)^2}$	$59_{(0,1,4,9)^2}$
$\mathbf{24}_{(0,2,5)^1}$	∞	∞	∞	∞	∞	∞	$100_{(0,2,5,7)^2}$	$108_{(0,2,5,8)^2}$	$51_{(0,2,5,9)^2}$
$\mathbf{15}_{(0,2,6)^1}$	∞	∞	∞	∞	∞	∞	$26_{(0,2,6,7)^2}$	$69_{(0,2,6,8)^2}$	$75_{(0,2,6,9)^2}$
$\mathbf{17}_{(0,1,7)^1}$	∞	∞	∞	$103_{(0,1,7,4)^2}$	$93_{(0,1,7,5)^2}$	$28_{(0,1,7,6)^2}$	∞	∞	∞
$\mathbf{63}_{(0,4,8)^1}$	$142_{(0,4,8,1)^2}$	$133_{(0,4,8,2)^2}$	$126_{(0,4,8,3)^2}$	∞	∞	∞	∞	∞	∞
$\mathbf{14}_{(0,2,9)^1}$	∞	∞	∞	$58_{(0,2,9,4)^2}$	$41_{(0,2,9,5)^2}$	$74_{(0,2,9,6)^2}$	∞	∞	∞
$\boldsymbol{w^*}_{(1,j)^2}$	$142_{(0,4,8,1)^2}$	$133_{(0,4,8,2)^2}$	$126_{(0,4,8,3)^2}$	$58_{(0,2,9,4)^2}$	$41_{(0,2,9,5)^2}$	$28_{(0,1,7,6)^2}$	$9_{(0,4,1,7)^2}$	$69_{(0,2,6,8)^2}$	$17_{(0,4,1,9)^2}$

Iteration 3: Exactly one node in every cluster has been visited. The salesman has to return to the origin node, thus modify node labels ${w^*}_{(1,k)^2}$ by connecting them back to the origin node to obtain a complete tour as shown in

Table 6.14 Partial tours after iteration 3.

	Origin	Cluster 1			Cluster 2			Cluster 3		
${w^*}_{(1,k)^2}$	$j=0$	$j=1$	$j=2$	$j=3$	$j=4$	$j=5$	$j=6$	$j=7$	$j=8$	$j=9$
$\mathbf{142}_{(0,4,8,1)^2}$	$154_{(0,4,8,1,0)^3}$	∞	∞	∞	∞	∞	∞	∞	∞	∞
$\mathbf{133}_{(0,4,8,2)^2}$	$136_{(0,4,8,2,0)^3}$	∞	∞	∞	∞	∞	∞	∞	∞	∞
$\mathbf{126}_{(0,4,8,3)^2}$	$149_{(0,4,8,3,0)^3}$	∞	∞	∞	∞	∞	∞	∞	∞	∞
$\mathbf{58}_{(0,2,9,4)^2}$	$59_{(0,2,9,4,0)^3}$	∞	∞	∞	∞	∞	∞	∞	∞	∞
$\mathbf{41}_{(0,2,9,5)^2}$	$46_{(0,2,9,5,0)^3}$	∞	∞	∞	∞	∞	∞	∞	∞	∞
$\mathbf{28}_{(0,1,7,6)^2}$	$60_{(0,1,7,6,0)^3}$	∞	∞	∞	∞	∞	∞	∞	∞	∞
$\mathbf{9}_{(0,4,1,7)^2}$	$65_{(0,4,1,7,0)^3}$	∞	∞	∞	∞	∞	∞	∞	∞	∞
$\mathbf{69}_{(0,2,6,8)^2}$	$160_{(0,2,6,8,0)^3}$	∞	∞	∞	∞	∞	∞	∞	∞	∞
$\mathbf{17}_{(0,4,1,9)^2}$	$101_{(0,4,1,9,0)^3}$	∞	∞	∞	∞	∞	∞	∞	∞	∞
${w^*}_{(1,j)^3}$	$46_{(0,2,9,5,0)^3}$	-	-	-	-	-	-	-	-	-

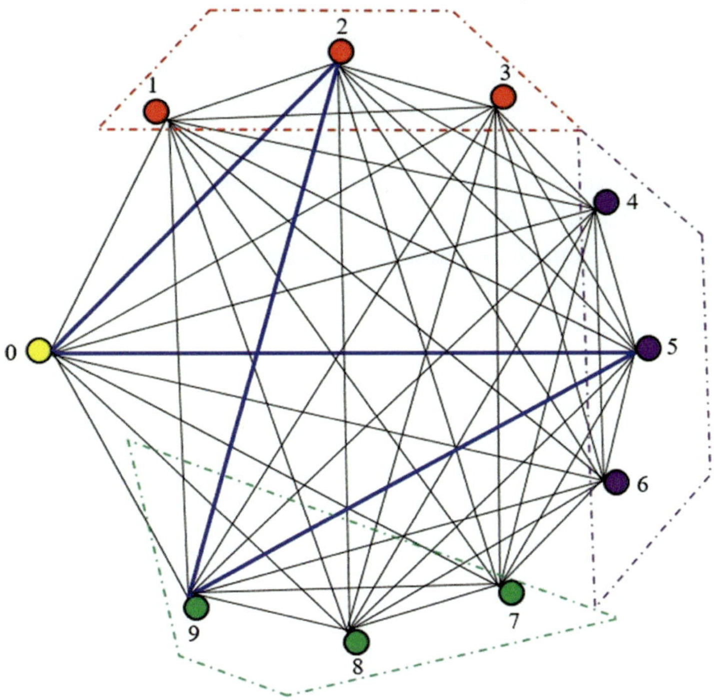

Figure 6.8 Solution to three clusters E-GTSP.

Table 6.14. We compute ${w^*}_{(1,j)^3}$ for the origin node. ${w^*}_{(1,j)^3}$ is the optimal solution to the equality generalized travelling salesman problem.

Minimum cost tour is given by 0→2→9→5→0 with tour distance of 46 units. Figure 6.8 shows the minimum tours on the E-GTSP network.

6.4 Concluding Remarks

In this chapter, travelling salesman network has been reconstructed using link weight modification based algorithm, namely the TANYAKUMU labelling algorithm. The algorithm was first presented by Tawanda et al. (2022). The algorithm is simple and computes optimal solutions always. The complexity of the algorithm reduces as the number of iterations increases. The algorithm requires reasonable number of iterations to compute the minimum cost tour in any given network.

References

Z.H. Ahmed, A comparative study of eight crossover operators for the maximum scatter travelling salesman problem. *International Journal of Advanced Computer Science and Applications*, *11*(6), 2020.

M.A.H. Akhand, S.I. Ayon, S.A. Shahriyar, N. Siddique, & H. Adeli, Discrete spider monkey optimization for travelling salesman problem. *Applied Soft Computing*, *86*, 105887, 2020.

A. Arigliano, T. Calogiuri, G. Ghiani, & E. Guerriero, A branch-and-bound algorithm for the time-dependent travelling salesman problem. *Networks*, *72*(3), 382-392, 2018.

E.O. Asani, A.E. Okeyinka, & A.A. Adebiyi, A construction tour technique for solving the travelling salesman problem based on convex hull and nearest neighbour heuristics. In *2020 International Conference in Mathematics, Computer Engineering and Computer Science (ICMCECS)* (pp. 1-4). IEEE, 2020, March.

N. Ascheuer, M. Fischetti & M. Grötschel, Solving the asymmetric travelling salesman problem with time windows by branch-and-cut. *Mathematical programming*, *90*, 475-506, 2001.

A. Asef-Vaziri, M. Kazemi, & M. Radman, The facility layout instances of the generalised travelling salesman problem. *International Journal of Production Research*, *60*(19), 5794-5811, 2022.

R. Bellman, Dynamic programming treatment of the travelling salesman problem. *Journal of the ACM (JACM)*, *9*(1), 61-63, 1962.

R. Bernardino, & A. Paias, Solving the family traveling salesman problem. *European Journal of Operational Research*, *267*(2), 453-466, 2018.

M. Boccia, A. Masone, A. Sforza, & C. Sterle, A column-and-row generation approach for the flying sidekick travelling salesman problem. *Transportation Research Part C: Emerging Technologies*, *124*, 102913, 2021.

U, Boryczka, & K. Szwarc, An effective hybrid harmony search for the asymmetric travelling salesman problem. *Engineering Optimization*, 2019.

G. Clímaco, L. Simonetti, & I. Rosseti, A branch-and-cut and MIP-based heuristics for the prize-collecting travelling salesman problem. *RAIRO-Operations Research*, *55*, S719-S726, 2021.

M. Dell'Amico, F. Maffioli, & P. Värbrand, On prize-collecting tours and the asymmetric travelling salesman problem. *International Transactions in Operational Research*, *2*(3), 297-308, 1995.

S. Fairee, C. Khompatraporn, S. Prom-on, & B. Sirinaovakul, Combinatorial artificial bee colony optimization with reinforcement learning updating for travelling salesman problem. In *2019 16th international conference on electrical engineering/electronics, computer, telecommunications and information technology (ecti-con)* (pp. 93-96). IEEE, 2019, July.

F.S. Gharehchopogh, & B. Abdollahzadeh, An efficient harris hawk optimization algorithm for solving the travelling salesman problem. *Cluster Computing*, *25*(3), 1981-2005, 2022.

H. Hernández-Pérez, & J.J. Salazar-González, A branch-and-cut algorithm for the split-demand one-commodity pickup-and-delivery travelling salesman problem. *European Journal of Operational Research*, *297*(2), 467-483, 2022.

A. Hossam, A. Bouzidi, & M.E. Riffi, Elephants Herding Optimization to Solve the Multiple Travelling Salesman Problem. In *Advanced Intelligent Systems for Sustainable Development (AI2SD'2019) Volume 4-Advanced Intelligent Systems for Applied Computing Sciences* (pp. 224-232). Springer International Publishing, 2020.

A.N. Júnior, & L.R. Guimarães, A greedy randomized adaptive search procedure application to solve the travelling salesman problem. *International Journal of Industrial Engineering and Management*, *10*(3), 238, 2019.

S. Kumar, E. Munapo, M. Lesaoana, & P. Nyamugure, A minimum spanning tree based heuristic for the travelling salesman tour. *Opsearch*, *55*, 150-164, 2018.

G. Laporte, & U. Palekar, Some applications of the clustered travelling salesman problem. *Journal of the operational Research Society*, *53*(9), 972-976, 2002.

P. Larranaga, C.M.H. Kuijpers, R.H. Murga, I. Inza, & S. Dizdarevic, Genetic algorithms for the travelling salesman problem: A review of representations and operators. *Artificial intelligence review*, *13*, 129-170, 1999.

K. Michalak, Evolutionary algorithm using random immigrants for the multiobjective travelling salesman problem. *Procedia Computer Science*, *192*, 1461-1470, 2021.

E. Munapo, A network branch and bound approach for the traveling salesman model. *South African Journal of Economic and Management Sciences*, *16*(1), 52-63, 2013.

M. Reda, A. Onsy, M.A. Elhosseini, A.Y. Haikal & M. Badawy, A discrete variant of cuckoo search algorithm to solve the Travelling Salesman Problem and path planning for autonomous trolley inside warehouse. *Knowledge-Based Systems*, *252*, 109290, 2022.

A. Santini, A. Viana, X. Klimentova, & J.P. Pedroso, The probabilistic travelling salesman problem with crowdsourcing. *Computers & Operations Research*, *142*, 105722, 2022.

R. Strange, A.Y. Yang, & L. Cheng, Discrete flower pollination algorithm for solving the symmetric travelling salesman problem. In *2019 IEEE Symposium Series on Computational Intelligence (SSCI)* (pp. 2130-2137). IEEE, 2019, December.

T. Tawanda, P. Nyamugure, S. Kumar, & E. Munapo, Modified TANYAKUMU Labelling Method to Solve Equality Generalized Travelling Salesman Problem. In *International Conference on Intelligent Computing & Optimization* (pp. 936-947). Cham: Springer International Publishing, 2022, October.

T. Tawanda, P. Nyamugure, S. Kumar, and E. Munapo, A Labelling Method for the Travelling Salesman Problem. *Applied Sciences*, *13*(11), p.6417, 2023.

T. Tóth, I. Detzky, & L. Eszes, Optimisation of Discrete Technology Processes Using A Method Traced Back to Constrained Travelling Salesman

Problem. In *Proceedings of the Twenty-Seventh International Matador Conference* (pp. 145-153). Macmillan Education UK, 1988.

B.D. Woods, & A.P. Punnen, A class of exponential neighbourhoods for the quadratic travelling salesman problem. *Journal of Combinatorial Optimization*, *40*(2), 303-332, 2020.

7

Some Reliability Routing Problems and Their Evaluation by Network Reconstruction

Abstract

In this chapter, stochastic networks are considered in the context of some reliability routing problems. After a brief description of a classical approach, we reconsider the maximum reliability route problem in a directed as well as in a non-directed network and solve them by the network reconstruction labelling approach. We have considered the terminal pair reliability and finally we considered protean stochastic networks and analysed them using the ring sum approach.

Keywords: Reconstruction in stochastic networks, maximum reliability and the maximum reliability route, maximum reliability route by dynamic programming, maximum reliability route by reconstruction of the link-probabilities, protean stochastic networks analysis, ring sum algebra.

Notations

1. $G(n, m)$ denotes the given network, (n) denotes the set of nodes in the network, (m) denotes the set of links in $G(n, m)$. These links indicate some relationship among various nodes i and j, where $i = 1, 2, \ldots, (n-1)$ and $j = 1, 2, \ldots, n$.
2. p_{ij} denotes a probability of successful communication, i.e., reliability of the link (i, j), where $0 \leq p_{ij} \leq 1$.
3. P_i = the maximum reliability of successfully communicating a message from a node (i) to the terminal node (N), through an optimal reliability route.

7.1 Introduction

In this chapter, we consider some stochastic networks in which links can fail or each link has an associated probably of successfully completing the intended task through that link. These probabilistic networks, known as stochastic networks, arise in various ways in real-life situations. Here we study them in the context of reliability routing problems. Stochastic networks have been considered by Ball (1979), Ball and Nemhauser (1979). The maximum reliability problem has been considered Bansal and Kumar (1977), Kumar (1979, 1981). Reliability routing problems were considered by Otway (1975), and by Petrovic and Jovanic (1979), Srivastava et al. (1972). Links joining the origin to the destination node denote a path in the network. Let us denote this path by S, where we are required to identify a path, denoted by S^*, where the path S^* has a property defined by the eqn (7.1), given below.

$$\{\text{Reliability of any path } S \leq \text{Reliability of the path } S^*\}. \tag{7.1}$$

This path denoted by S^* is known as the maximum reliability path. The problem is to identify the path as well as the maximum reliability. Maximum reliability was considered by Srivastava et al. (1972), where they used the dynamic programming approach. In that paper, the authors have evaluated the maximum reliability and the associated path with that reliability. After describing the dynamic programming approach, we reconsider the same problem using the reconstruction concept. This reconstruction idea in context of a shortest path in a directed network was developed by Munapo, et al. (2008) and for the non-directed network, it was developed by Kumar et al. (2013). This chapter has been organized in the following sections:

In Section 7.2, maximum reliability route has been determined by using the dynamic programming approach due to Bellman and Dreyfus (1962), which was used by Srivastava et al. (1972).

In Section 7.3, we describe a single element label and the reconstruction approach. In Section 7.4, we discuss the labelling approach using the single element labelling approach. We also give some numerical illustrations. The terminal pair reliability problem has been discussed in Section 7.5, stochastic protean networks have been discussed in Section 7.6 and finally the chapter has been concluded in Section 7.7

7.2 Maximum Reliability Route in Communication Networks by the Dynamic Programming Approach

Communication networks have experienced a great deal of technological advances; however, in many situations, communication from a station (i) to another station (j) remains uncertain, therefore communication through a link can be represented by a probability, and the problem is to find the maximum reliability and the associated maximum reliability route in a stochastic communication network from a desired origin node to a destination node.

Srivastava et al. (1972) considered this problem, and defined:

P_i = the maximum reliability of successfully communicating a message from a node (i) to the terminal node (N), through an optimal reliability route.

p_{ij} = the probability of successfully communicating through the link (i, j).

These two probabilities can give rise to a recurrence relation in eqn (7.2)

$$P_i = \max\left[p_{ij} X P_j\right], \; i = 1,\ 2,\ \ldots,\ N-1 \; with \; P_N = 1. \tag{7.2}$$

Eqn (7.2) can be solved recursively with the help of the boundary condition $P_N = 1$ and one more conditional probability defined below and used in eqn (7.3).

The successive approximation can be computed with the help of the boundary condition and the maximum reliability route can be obtained. Since we do not know the number of links required by the maximum reliability route, we will calculate all possibilities, i.e., direct connection (or one link connection), two link connection, etc., and pick from all possibilities the maximum reliability route. This approach is computationally heavy. For using this approach, we define one more conditional function and develop another relation given by eqn (7.3).

P_i^K = the maximum reliability of successfully communicating a message from node i to the terminal node N through a path which consists of K number of distinct links. This results in eqn (7.3)

$$P_i^{K+1} = \max j\left[P_{ij} X P_j^K,\ i \neq j\right].$$

With

$$P_i^1 = P_{iN}; (i = 1, 2, \ldots,\ N-1), P_N = 1. \tag{7.3}$$

These computations, which can be demanding for larger networks, are illustrated in Section 7.2.1.

7.2.1 A numerical illustration

Consider a stochastic network taken from Srivastava et al. (1972); it is given in Figure 7.1 along with the associated probabilities.

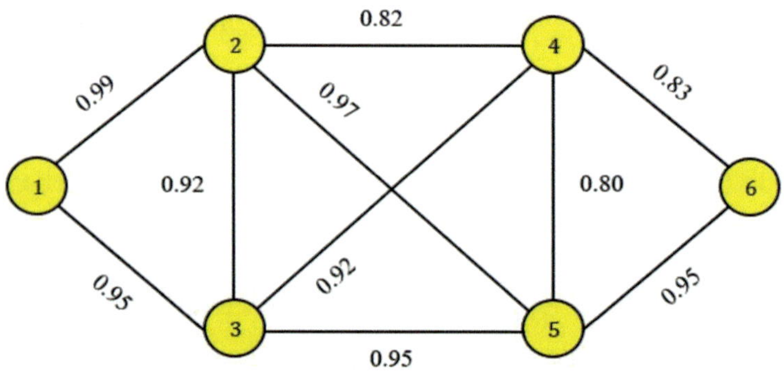

Figure 7.1 Reliability network for an illustration.

These link probabilities are also given in Table 7.1.

Table 7.1 Link probabilities.

From/to	1	2	3	4	5	6
1	1	0.99	0.95	0	0	0
2	0.99	1	0.92	0.82	0.97	0
3	0.95	0.92	1	0.92	0.95	0
4	0	0.82	0.92	1	0.80	0.83
5	0	0.97	0.95	0.80	1	0.95
6	0	0	0	0	0	1

The zeros in row 6 in Table 7.1 indicate that once reached node 6, it is unnecessary to leave that node. From the last column of Table 7.1, we have P_i^1 values, as given in Table 7.2.

Table 7.2 Values of P_i^1.

$K\backslash i$	1	2	3	4	5	6
1	0	0	0	0.83	0.95	0

These values of P_i^1 and probabilities in Table 7.1 will give us the value of P_i^2, as shown in Table 7.3.

Table 7.3 Values of P_i^2.

$i\backslash j$	**1**	**2**	**3**	**4**	**5**	**6**	P_i^2
1	0	0	0	0	0	0	0
2	0	0	0	0.6806*1	0.9215	0	0.9215[5]#
3	0	0	0	0.7636*2	0.9025	0	0.9025[5]
4	0	0	0	0.8300**	0.7600	0	0.7600[5]
5	0	0	0	0.6640	0.9500**	0	0.6640[4]
6	0	0	0	0	0	0	0

Note: Illustration of values calculated in Table 7.3, for example see *1 and *2 values, they have been obtained $P_{24} \times P_4^1 = 0.82 \times 0.83 = 0.6806$, and $P_{34} X P_4^1 = 0.92 \times 0.83 = 0.7636$ and the values with double asterisk indicate that value is inconsistent with the definition. Nodes [5] and [4] indicated in the last column give details of the two-link route, for example in the last column in the row for node 2 is 0.9215[5] indicates that the two-link route is from node 2 to 6 and that path is from node 2 to node 5 and from node 5 to node 6 with joint reliability 0.9215.

Similarly, we can calculate P_i^3, P_i^4 and P_i^5. In a 6-node network, no path can have more than 5 links, hence calculation process can be terminated. This summary is given in Table 7.4.

Table 7.4 Summary of P_i^k values (* indicates an optimal value).

$k\backslash i$	**1**	**2**	**3**	**4**	**5**
1	0	0	0	0.8300	0.9500
2	0	0.9215	0.9025	0.7600	0.6640
3	0.912285*	0.8303	0.8478	0.8303	0.8939
4	0.8220	0.9032	0.8667	0.7800	0.8454
5	0.8942	0.8671	0.8492	0.7974	0.8761

Since we are interested in the maximum reliability path from node 1 to node 6, by inspection, we find from column 1 for $i = 1$, the maximum value is 0.912285 for $k = 3$. It means that the path is comprised of 3 links. The path is:

$$1 \to 2 \to 5 \to 6 \text{ with probabilities } = 0.99 \times 0.97 \times 0.95 = 0.912285. \tag{7.4}$$

7.2.2 Comments and observations

Observation 1:

In the above illustration, we are dealing with probabilities, i.e., numbers between 0 and 1. If we denote probability of two links (i_1, j_1) and (j_1, j_2) by two numbers ‘a’ and ‘b’, which are between 0 and 1, the joint probability of the path consists of two links $(i_1 \rightarrow \; j_1 \rightarrow \; j_2 \;)$ will be given by the product, $a.b \; \leq \min{(a, b)}$. This can be easily established. For example, let the $\min{(a, b)}$ be a.

Now evaluate the function:

$$a - a.b = a\,(1 - b) \geq 0. \tag{7.5}$$

Hence the product is less than the minimum of the two. Also note that when the larger quantity is equal to 1, i.e., perfect or the reliability is 1, then the RHS is zero, indicating the product will be equal to the smaller of the two given probabilities or difference will be zero as shown by eqn (7.5).

Observation 2:

From observation 1, it can be said that once a path made by three links was identified, subsequent paths made of four or five links are less likely to have probability more than a path comprised of three links. This is clearly displayed in the summary in Table 7.4. However, it is all a function of the given reliabilities, see the numerical illustration shown in Section 7.3.2.

7.3 Reconstruction of a Probabilistic Network and Development of a Labelling Approach for the Maximum Reliability Path

For the application of the reconstruction approach, we need to define a label and purpose of that babel associated with a node in a reliability network.

7.3.1 What is a label?

Kumar et al. (2016) determined a maximum reliability path from the origin node to that node by developing a two-element label. These two elements represented the reliability of that path (i.e., the reliability of the path from the origin node to that node), and the second element represented the node number to identify the node from where we arrive to the current location. These two elements are denoted by (r, i), where *r* denotes the reliability of the path from the origin node to that node and *i* denotes the node from where we arrive to the current node.

7.3.2 An initial label in a network

Fortunately, in each stochastic network a node can be identified to initiate labelling computations for finding the maximum reliability path. This label is $(1,1)_1$ indicating that we reach the origin node from itself and the suffix 1 indicates that it is the first label, the reliability is 1 and the path is $1 \rightarrow 1$. The suffix is used to indicate the label number, which at the origin node is 1 in this situation. Therefore, it is justified to assume that the origin node is a labelled node for any given stochastic network.

7.3.3 Property of a labelled node

A labelled node gives the maximum reliability from the origin node to that node; hence we will never re-enter a labelled node from any other node. This can be easily seen by the fact that has been discussed in Section 7.2.2, which states that the combined reliability is less than the minimum of the individual reliabilities of links that constitute a path.

7.3.4 Development of a new one-element label: A reconstruction idea

We explain the reconstruction idea with the help of a typical network shown in Figure 7.2. The origin node has a label (1,1), indicating that the node is perfect, hence reliability 1, and the path is we reached here from this node itself. Now consider that three links are going out from this origin node are $(0, l), (0, m)$, and $(0, n)$. Let the associated reliabilities of three links be : $(r_{0l}, r_{0m},\ r_{0n}) = (0.91, 0.95, 0.93)$. The maximum reliability is 0.95 associated with the link (0,*m*). Furthermore, let the reliability associated with three links going out from node *m* be denoted by $(r_{mo}, r_{mp},\ r_{mq})$. The idea of reconstruction is to modify these reliabilities without changing the overall reliability of the maximum reliability path from the origin node to the destination node. Since the maximum reliability value 0.95 is associated with the link (o,*m*), denoted by r_{om}. Let this maximum reliability of the link (O,*m*) be changed to 1, i.e., made it perfect. The three links going out of node *m* be denoted by:

$(r_{mo}, r_{mp},\ r_{mq}) = (0.92,\ 0.96,\ 0.94)$ can be changed by multiplying them by 0.95 and altered as: $(0.95 \times 0.92, 0.95 \times 0.96, 0.95 \times 0.94) = (0.874, 0.912, 0.893)$, which is not altering the reliability $(0m \rightarrow m0 \text{ or } 0m \rightarrow mp \text{ or } 0m \rightarrow mq)$. The reconstructed reliabilities for four links will be given by: $(r_{om}, r_{mo}, r_{mp}, r_{mq})$ = (1.0, 0.874, 0.912, 0.893). Similar process of labelling will continue, until the destination node is labelled. These perfect links, for example, help us to identify the maximum reliability route to reach the labelled node to the starting node 1. Therefore, in the proposed reconstruction approach, only one label, indicating the reliability, is sufficient.

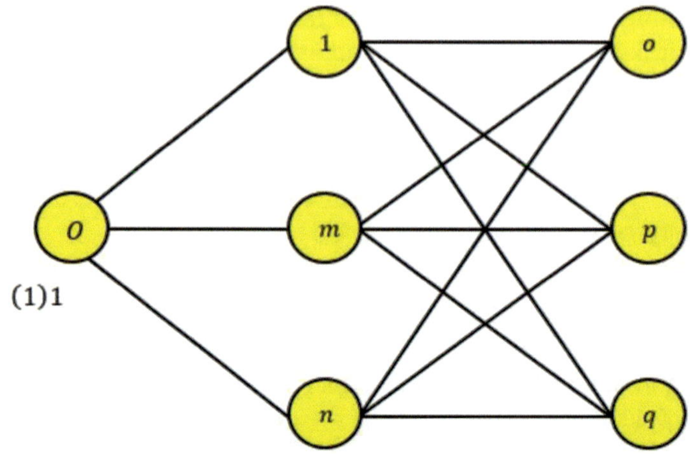

Figure 7.2 A typical network.

7.4 Reliability Evaluation by Network Reconstruction in a Probabilistic Network

Consider a directed network $G(n, m)$, where nodes can be numbered such that the network links represented by (i, j), are such that $i < j, i, j = 1, 2, \ldots, n$ for all links (i, j). It means that nodes are numbered according to a topological sort, which is possible for a directed network. We are assuming nodes are perfect, i.e., the reliability associated with all nodes is 1, however, link (i, j) is associated with a probability, p_{ij} such that $0 < p_{ij} < 1$. We are required to find the maximum reliability route, which can be obtained by the following steps. Directions were useful for the conventional shortest route problem, however, for the maximum reliability path, there is no such advantage. Consider the following steps:

Step 1: The source node is denoted by node 1 and initially this node 1 can be labelled (1), indicating that the reliability is to reach node 1 is 1. In the reconstruction approach only one element label is sufficient, as the path associated with the maximum reliability can be identified by the link with reconstructed reliability 1(indicating a fully reliable link). This one element label reflects the reliability from the origin node to that node. For labelling, we use the notation *L*; thus the label at node 1 will be *L*(1).

Step 2: Set the labelled nodes as nodes $k_1, k_2, \ldots, k_r$, where r is the number of labelled nodes at any stage. Initially $r = 1$, i.e., the origin node, which is the only labelled node.

Step 3: From node *k*, find the link with a maximum outgoing reliability, which can be obtained $p_k = \max\{p_{k,1,}; p_{k2}; \ldots, p_{kr}\}$, assuming r represents the number of outgoing links from node *k*. Let p_{kl} is the maximum reliability value associated with the links (*k*,*l*) outgoing from the labelled node *k*. In case of a tie, resolve the tie arbitrarily. The probability of the maximum reliability link denoted by p_k is altered to perfect link, i.e., 1 and reliability of the outgoing links (*j*,*q*) from node *j* is modified by multiplying their reliabilities by the reliability p_k.

Step 4: If the destination node is not labelled, move to the next iteration, i.e., $k = k + 1$ and return to step 3.

Step 5: When the destination node '*n*' is labelled, that becomes the maximum reliability of the path joining the origin node to the destination node. The route joining the origin node to the destination can be back traced by moving through the perfect links backwards towards the origin node.

This algorithm can now be illustrated by a numerical illustration.

7.4.1 Numerical illustration for a directed network

Let us reconsider a minor variation of the directed network taken from Kumar et al. (2016) shown in Figure 7.3. and find the maximum reliability and the maximum reliability route by using the network reconstruction approach.

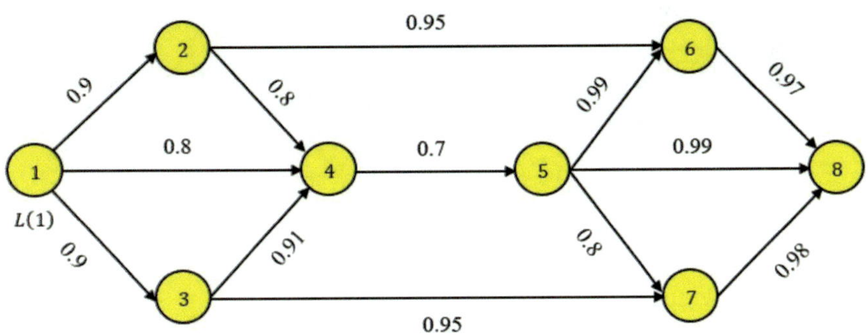

Figure 7.3 Directed network illustration.

Node 1 is the labelled node initially, and the label is *L*(1), as the node is perfect. From this node, there are three outgoing links and maximum reliability of these three links is given by 0.9, which corresponds to links (1,2) and (1,3). We arbitrarily select, say node 2, and make reliability adjustments as shown in Figure 7.4. These adjustments will be for the links (1,2), (2,4) and (2,6).

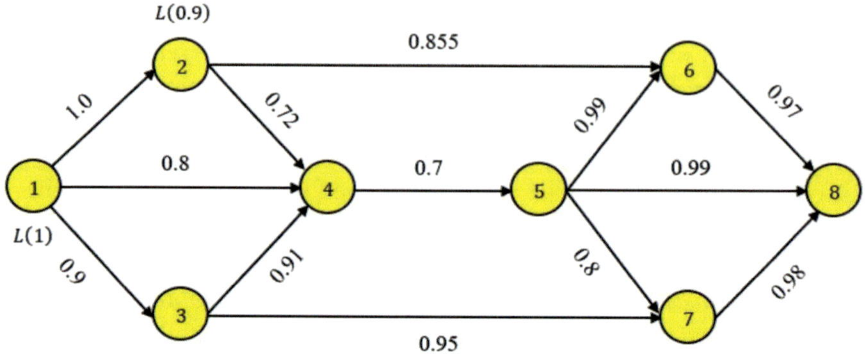

Figure 7.4 The network after one labelling iteration.

There are four outgoing links from the labelled nodes, i.e., two links from node 1 and two links from node 2. These four links are: (1,3), (1,4), (2,4) and (2,6). The maximum reliability is once again 0.9 that corresponds to node 3, hence the next node to be labelled will be node 3. It will alter only the link reliabilities of the links (1,3), (3,4) and (37). These new reliabilities will be 1 for (1,3), and 0.819 for the link (3,4) and 0.855 for the link (3,7). This information is upgraded in Figure 7.5.

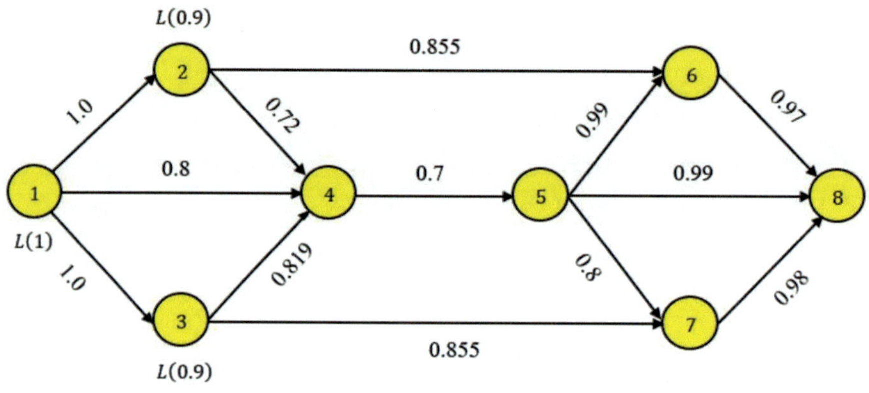

Figure 7.5 Updated network after labelling node 3.

Now there are several outgoing links from these three labelled nodes, which are nodes 1, 2 and 3. These links are (1,4) (2,6), (2,4), (3,4) and (3,7). The maximum corresponds to 0.855 for the links (2,6) and the link (3,7). Let us label node 6 and the updated network is shown in Figure 7.6.

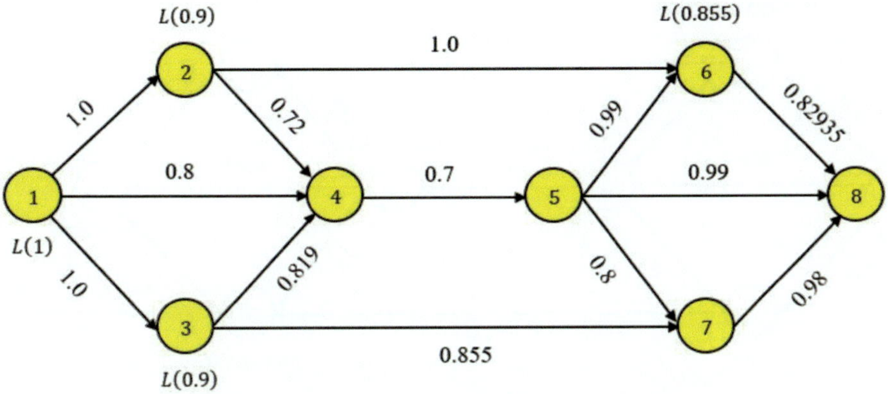

Figure 7.6 Updated network after labelling node 6.

Note nodes 4 and 5 have not been labelled and they will not be labelled, as the maximum reliability link is (3,7), which will lead to the next labelled node, which is node 7. Also note that the label on node 6 has altered only the reliability of the link (2,6), which is now perfect, and the reliability of the link (6,8), which is 0.82935. This is shown in Figure 7.7 along with updated link reliability information. Next node to be labelled will be node 7.

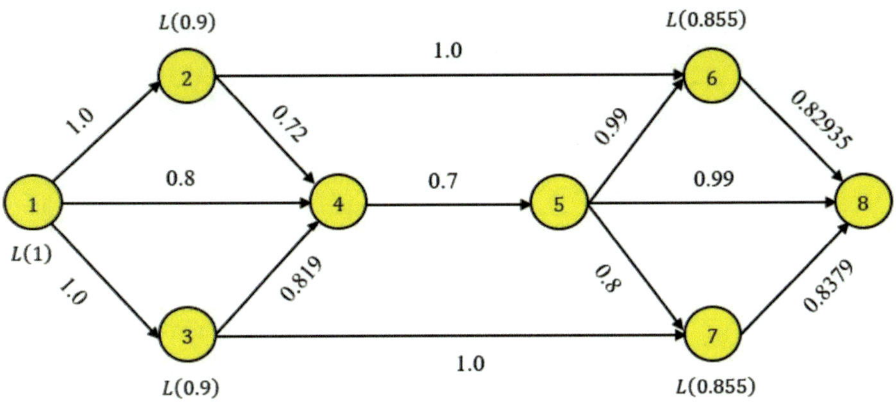

Figure 7.7 The updated network after labelling node 7.

Hence the maximum reliability will be the max (0.82935, 0.8379) = 0.8379. The label on the destination node will be $L(0.8379)$ and the updated network will be as given in Figure 7.8.

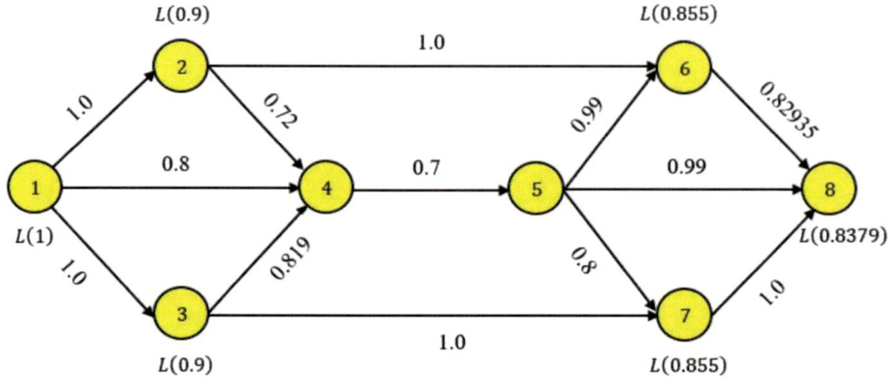

Figure 7.8 Label on node 8 indicates the maximum reliability.

The maximum reliability route can be easily traced back with reconstructed link reliabilities, i.e., $8 \leftarrow 7 \leftarrow 3 \leftarrow 1$ or the maximum reliability path will be given by: $1 \rightarrow 3 \rightarrow 7 \rightarrow 8$.

7.4.2 Numerical illustration: Non-directed network

Let us reconsider the same network without the directions, i.e., we consider it as a non-directed network, as shown in Figure 7.9.

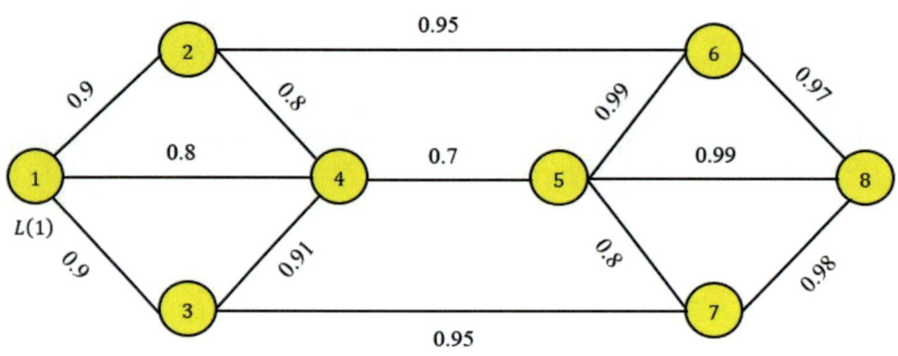

Figure 7.9 Probabilistic non-directed network.

Once again, the network in Figure 7.9 is the same as was considered with directions in Figure 7.3. The maximum reliability associated with the outgoing links from node 1 is 0.9 associated with two links (1,2) and (1,3). After labelling node 1 as $L(1)$, suppose we select arbitrarily node 2, the link reliability of the link (1,2) is changed from 0.9 to 1 (i.e., it is made perfect) and the two outgoing links (2,6) and (2,4) will become $0.95 \times 0.9 = 0.855$ and $0.8 \times 0.9 = 0.72$. Similarly, the labels will repeat on nodes 2, 3, 6. This is shown in Figure 7.10. Note the label on node 6 has altered the reliability of the link (6,5) as this link is a non-directed link. It was not the case in the directed network as the link (5,6) was directed from node 5 towards node 6.

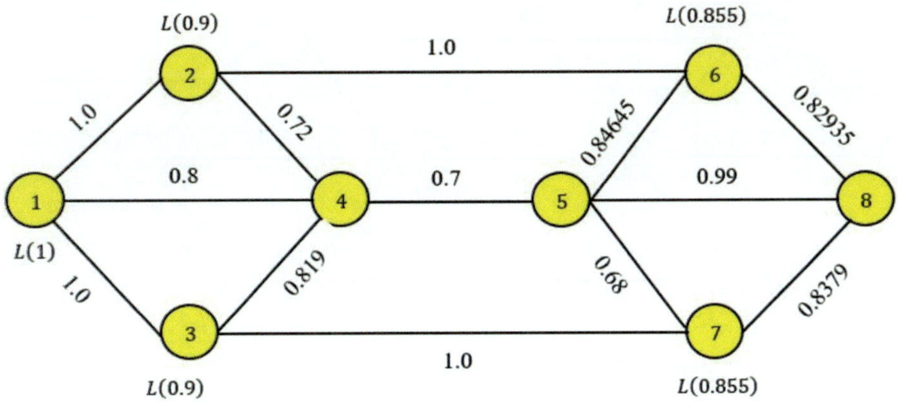

Figure 7.10 Reconstructed network after the first label on node 1.

In the network in Figure 7.10, there are several labelled nodes, i.e., nodes 1, 2, 3, 6, 7. Therefore, we consider, all links going out from these labelled nodes, which are: (1,4), (2,4), (3,4), (6,8), (6,5), (7,5) and (7,8). Reliabilities of these links are: $\{0.8, 0.72, 0.819, 0.82935, 0.84645, 0.68, 0.8379\}$.

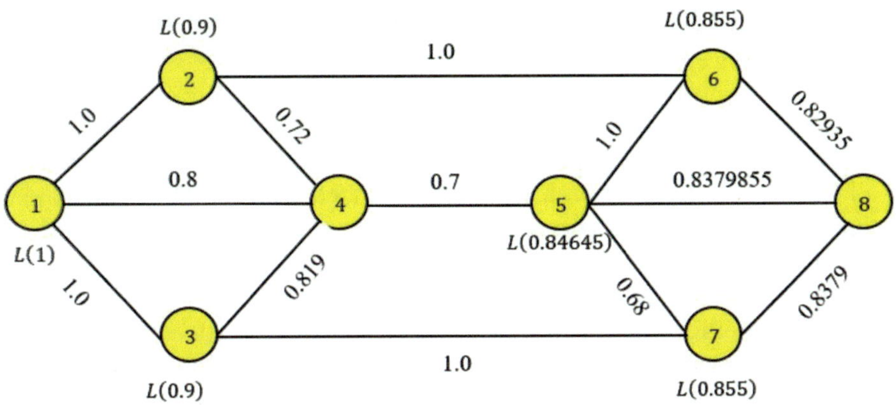

Figure 7.11 Updated network after labelling node 5.

Maximum among them is 0.84645 along the route (6,5). Therefore, node will be the next node to be labelled is node 5 and the label will be L(0.84645). This is shown in Figure 7.11.

Finally, the label on node 8 will be the maximum of $\{0.82935, 0.8379855, 0.8379\} = 0.8379855$. See Figure 7.12.

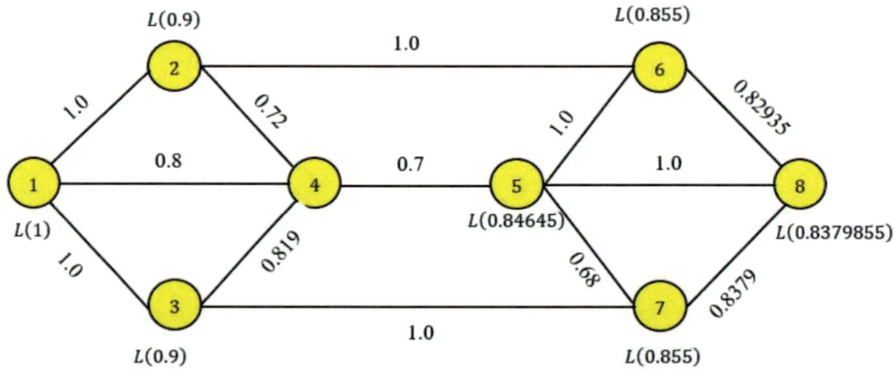

Figure 7.12 Network after the label on node 8.

The maximum reliability is 0.8379855 and the maximum reliability path by back tracing is given by:$1 \rightarrow 2 \rightarrow 6 \rightarrow 5 \rightarrow 8$.

7.5 Terminal Pair Reliability in a Reliability Network

Once again, we consider a directed network $G(n, m)$, where n represents the set of nodes $\{1, 2, \ldots, N\}$ and m represents the set of links joining various nodes. In the directed network, nodes can be numbered such that the network links represented by (i, j), are such that $i < j,\ i,\ j = 1, 2, \ldots, n$ for all links (i, j). It means that nodes are numbered according to a topological sort, which is possible only for the directed network where for each link, $i < j$. We are assuming nodes are perfect, i.e., the reliability associated with all nodes is 1, however, link (i, j) is associated with a probability, p_{ij} such that $0 < p_{ij} < 1$. In a non-directed network, the condition $i < j$ will not be satisfied as communication from i to j is allowed and also from j to i. An important measure of stochastic network is known as the 'terminal pair reliability' (TPR) which is the probability that there exists at least one path from a specified origin node 1 to another specified terminal node N, along with all nodes and links are in perfect state, that is, they are in the working state. The value of TPR has been also evaluated by Fratta and

Montanari (1973), Hansler et al. (1974), Kim, Case and Ghare (1972), and Kumar (1982). It is assumed that state of each component is in one of the two states – good or bad, i.e., quantitative classification is not possible. TPR calculations are simple for a series-parallel network, else they become very complex, see Lin et al. (1976), where it has been estimated to be $2^{(m-1)}$, where m represents the number of path-sets in the given network. Fretta and Montanari (1973) used the Boolean approach, Kumar (1982) used the dynamic programming along with the Boolean approach, which is used in this chapter.

7.5.1 Assumptions and notation

7.5.1.1 Assumptions

1. The nodes are perfect, i.e., they cannot fail.
2. Link reliabilities are independent of time.
3. The state of each link is statistically independent or in short, S-independent.

7.5.1.2 Notation

1. $i,\ j = 1, 2, \ldots,\ N$, where 1 and N are the terminal nodes.
2. (i, j) is a link joining the nodes (i and j).
3. R = the terminal pair reliability between nodes 1 and N, which is to be determined.
4. p_{ij} = the reliability associated with the link $(1, j)$.
5. f_i = the terminal pair reliability between the nodes, i and the fixed terminal node N for an acyclic directed network, where $f_1 = R$ (to be determined).
6. f_i^n = the terminal pair reliability between nodes i and node N for a non-directed network when the paths are restricted to have at most n links; the value to be determined $R = f_1$.
7. A_i = the set of links going out from the node i, i.e., immediate successor nodes of node i.
8. The * operation reduces the exponents to 1, for example the expression $\left[p_{23}^2 \cdot p_{34}^3 p_{12} + p_{23}^3 p_{12}^4\right] = p_{23} \cdot p_{34} \cdot p_{12} + p_{23} p_{12}$, since the same link may appear in several paths.

7.5.2 Mathematical equation for the directed acyclic network

In a directed acyclic network, it is possible to number the nodes in such a way that if there exists a link directed from node i to node j, then $i < j$. In such a network, it can be seen that f_i and p_{ij} can be connected to form the following dynamic programming recurrence relation, given by eqn (7.6):

$$f_i = \left[1 - \left\{ \prod_{j \in A_i} (1 - p_{ij} f_j) \right\} * \right], \ i = N-1,\ N-2,\ \ldots,\ 1; j > i. \quad (7.6)$$

Eqn (7.6) can be evaluated iteratively with the condition given by eqn (7.7)

$$f_N = 1, \text{ as the node is perfect.} \quad (7.7)$$

In the Section 7.5.3, we illustrate computations involved in the recurrence relation in eqn (7.6).

Note that for performing the * algebra, we must retain algebraic form of various functions, which can soon become unwieldy, therefore we will for our use evaluate them and assign a numerical value for subsequent calculations. This will give approximate estimates of the terminal pair reliability.

7.5.3 A Numerical illustration for the relation

Consider the directed network in Figure 7.13.

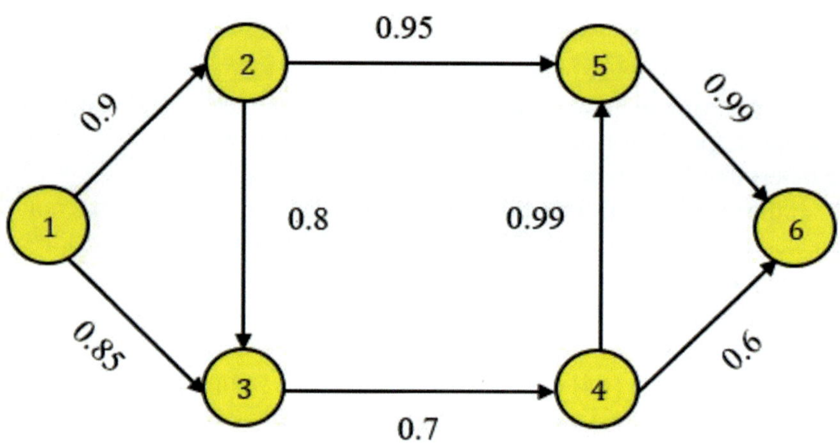

Figure 7.13 Directed network for the terminal pair reliability.

From eqn (7.7), we have $f_6 = 1$, the value of $N = 6$. Using eqn (7.6) for $N - 1$, i.e., for 5, $f_5 = p_{56} = 0.99$.

Similarly for $N-2=4$, we have:

$$\begin{aligned} f_4 &= 1 - \left[\prod_{j\in A_4}(1-p_{4j}f_j)\right]^* \\ &= 1-[(1-p_{45}f_5)(1-p_{46}f_6)]^* \\ &= [p_{45}f_5+p_{46}f_6-p_{45}p_{46}f_5f_6]^* \\ &= [p_{45}p_{56}+p_{46}f_6-p_{45}p_{46}p_{56}f_6]^* \\ &= [p_{45}p_{56}+p_{46}-p_{45}p_{46}p_{56}]^* \\ &= 0.99\times 0.99+0.6-0.99\times 0.6\times 0.99 = 0.99204. \end{aligned}$$

Once f_4, f_5, and f_6 are known, we can use eqn (7.6) to find the next value of f_3, which is given by:

$$f_3 = p_{34}.f_4 = p_{34}[p_{45}p_{56}+p_{46}-p_{45}p_{46}p_{56}]^*.$$

Note exponent will have no effect as the expression in the square brackets is free of p_{34}. Hence, we can evaluate the value of f_3, which is given by:

$$f_3 = 0.7\times\ 0.99204 = 0.694428.$$

We can now calculate the value of:

$$\begin{aligned} f_2 &= 1 - \left[\prod_{j\in A_2}(1-p_{2j}f_j)\right]^* \\ &= 1-[(1-p_{23}f_3)(1-p_{25}f_5)]^* \\ &= [p_{23}f_3+p_{25}f_5-p_{23}p_{25}f_3f_5]^* \\ &= p_{23}f_3+p_{25}f_5-[p_{23}p_{25}\times 0.694428\times 0.99]^* \\ &= 0.8\times 0.694428+0.95\ \times\ 0.99-[0.8\times 0.95\times 0.694428\ \times\ 0.99\,] \\ &= 0.5555424+0.9405--0.52248763 = 0.97355477. \end{aligned}$$

Finally, we have from eqn (7.6) the required terminal pair reliability given by:

$$f_1 = 1 - \left[\prod_{j\in A_1}(1-p_{1j}f_j)\right]^*$$

$$
\begin{aligned}
&= 1 - [(1 - p_{12}f_2)(1 - p_{13}f_3)]^* \\
&= [p_{12}f_2 + p_{13}f_3 - p_{12}p_{13}f_2f_3]^* \\
&= 0.9 \times 0.97355477 + 0.85 \times 0.694428 - 0.9 \times 0.85 \times 0.97355477 \\
&\qquad \times 0.694428 \\
&= 0.87619929 + 0.5902638 - 0.51718872 = 0.94927437.
\end{aligned}
$$

It means that the terminal pair reliability between the origin node 1 and the terminal node 6 is estimated by 0.94927.

This method results in an approximate estimation of the terminal pair reliability. For exact value, we must hold on to the expressions in the algebraic form and carry out the * operation for each intermediate function values.

7.6 Stochastic Protean Networks: Alternative Paths

A system that changes its state while performing the intended task is described as a protean system, see Kumar (1995). Changes in a system may arise within that system or system may be influenced due to some external interactions, which are generally beyond the control of that system. In cases when a system changes its state, it is desirable for a good system to be able to continue discharging its intended responsibilities. In many systems, there are only two states, an up state, i.e., when it is working, and the other one is the down state, when it is not working. For example, one such system of interest arises in the telecommunication network, where due to failure of some of the communication links, a usual path between the two points of interest; communication may be disrupted due to failure of some links on its usual path. However, if the communication goes through some alternative links in the up state, the end user does will not face any disruption. This is more so because high capacity of information can flow through the modern communication links. Each link is much more important, as in the case of a disruption, an alternative path that can fill the place can avoid that disruption. This requirement is more than ever before, see Acampora and Karol (1989), and Herzberg and Bye (1994).

When some links have been identified in the down state, the problem is to find a path that does not pass through the failed or forbidden links. It is a reverse problem to a problem that requires a path that passes through the set of specified edges, see Kalaba (1960), and Kumar and Arora (1995), or Kumar et al. (1999). Therefore, the problem considered in this section is similar but inverse, where we are required to find paths which does not pass through a set of edges which does not pass through the set of forbidden edges.

7.6.1 Notation used in this section

Consider a directed network $G(n, m)$, where n denotes the set of nodes ($n = 1, 2, \ldots, N$) and m denotes the set of links such that $m = \{l_1, l_2, \ldots, l_m\}$. These links are connecting the n nodes. It is assumed that in the normal state of the network system, between every pair of nodes, there is a path. Consider one such path P between a given pair of nodes and assume that this path is made by y number of links. When some of these y links are in the failed state, the path P between the same pair of nodes is in the down state. Therefore, establishment of an alternative path between the same pair of nodes becomes necessary. It means we are looking for an alternative path joining the same pair of nodes but does not pass through any of the y failed edges.

Let the set of failed edges be denoted by the set $X = \{a_1,\ a_2,\ \ldots,\ a_x\}$. Note that the set X is a subset of the set Y. Further, this alternative path should not be too long. In other words, we need the alternative path made of at most K number of links. This integer K has a preassigned known value.

Circuits and Euler graphs which are subgraphs of the given graph are represented by C and E respectively. A subgraph, which is not a path, is represented by Q. Different circuits, Euler graphs, paths, and subgraphs (which are not paths) are represented by $C_i,\ E_i,\ P_i,$ and Q_i respectively.

A few terms are necessary for development of the rest of the model.

1. Fundamental circuits: In a n node network, a connected tree t will be made of $(n-1)$ links, the number of chords will be $(m-(n-1)) = (m-n+1)$. Let $\{c_i\}$ be a circuit made by the $t \cup c_i\ for\ i = 1,\ 2,\ \ldots,\ (m-n+1)$. The collection of circuits $\{c_1,\ c_2,\ \ldots, c_{m-n+1}\}$ is called a set of fundamental circuits with respect to the tree t.
2. The ring sum of two sets A and B is given by : $\mathrm{A} \bigoplus \mathrm{B} = \mathrm{A} \bigcup \mathrm{B} - \mathrm{A} \bigcap \mathrm{B}$.
3. The number of elements in the ring sum $\{A \bigoplus B\}$ is given by $|A \bigoplus B|$, which can be obtained by the RHS of the equation is given by:

$$\left\lceil A \bigoplus B \right\rceil = |A| + |B| - 2\,|A \cap B|\,. \tag{7.8}$$

4. The set of fundamental circuits C can be partitioned into three disjoint subsets denoted by $C_N,\ C_S,$ and C_A, and they represent sets which contain no failed unit from the set X, contain some of the failed units in the set and contain all failed units of the set X, respectively. Let the number of such circuits be denoted by the number $a,\ b,$ and c. In other words, it means that: $C_N = \{C_{N1},\ C_{N2},\ \ldots,\ C_{Na}\}$.

5. Therefore $a = |C_N|$. Similarly, we have: $C_S = \{C_{S1},\ C_{S2},\ ...,\ C_{Sb}\}$ *or* $b = |C_S|$ and $C_A = \{C_{A1},\ C_{A2},\ ...,\ C_{Ac}\}$ or $c = |C_A|$.
6. From items 4 and 5 above, it is obvious that $a + b + c = f$. It may be noted that partitioning concept is essential for the protean networks, for example, the set of failed edges may change at any subsequent time due to failure of some other link or due repair of the failed link. As the set X changes, the number a, b,and c will change to a', b', and c' but $a' + b' + c' = f$.
7. *P* is the original path connecting the pair of nodes under consideration, which passes through the specified links in the failed state. This set of failed links was denoted by the set X.

7.6.2 The algorithm to find the alternative path free of the failed edges.

For the given network, we assume that the following is known.

1. The path *P* joining the two nodes and passing through the set of failed edges; hence the current path is in the down state.
2. The shortest connected network of the given graph, chords, and the fundamental circuits.
3. The set X is known.
4. The partitioned sets with respect to the elements of the set X, i.e., the sets C_N, C_S, and C_A.

We now illustrate the calculations involved in determination of the alternative path.

7.6.3 Numerical illustration

Consider a linear graph in Figure 7.14 comprised of 11 nodes and 15 links joining various nodes. The shortest connected graph will be comprised of 11 nodes and 10 links. Hence this network will have 5 chords and 5 fundamental circuits.

Let the path *P* between nodes 1 and 11 be given as: $1 \rightarrow 2 \rightarrow 3 \rightarrow 4 \rightarrow 9 \rightarrow 11$, which is having three links (1,2), (2,3) and (3,4) in the failed state. Hence, we are looking for an alternative path joining nodes 1 and 11 and it does not contain the failed links i.e., (1,2), (2,3) and (3,4). In this illustration, we have five fundamental circuits with respect to each chord shown in red. These circuits are shown in Figure 7.15.1–7.15.4.

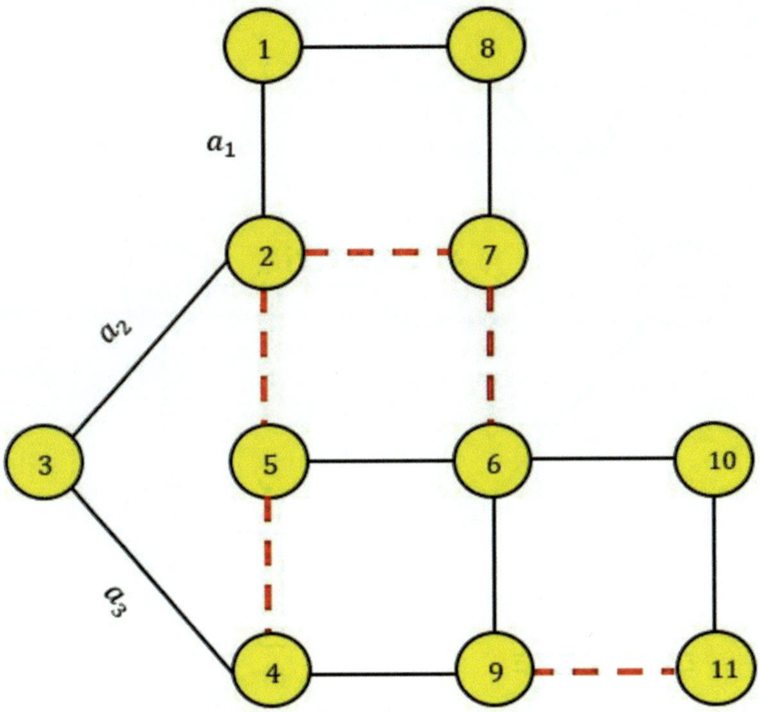

Figure 7.14 Graph for an alternative path.

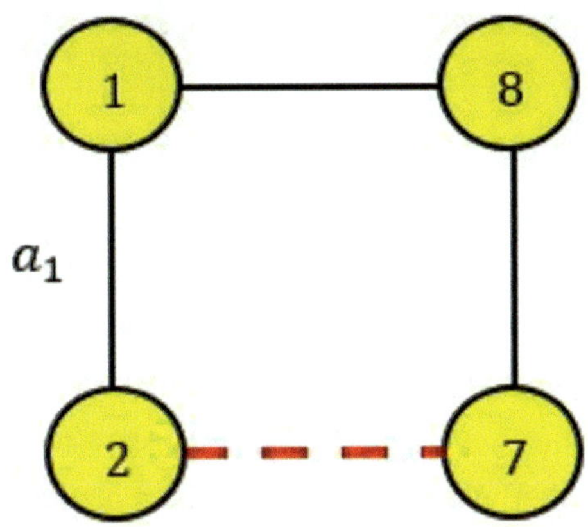

Figure 7.15.1 Fundamental circuit C_1.

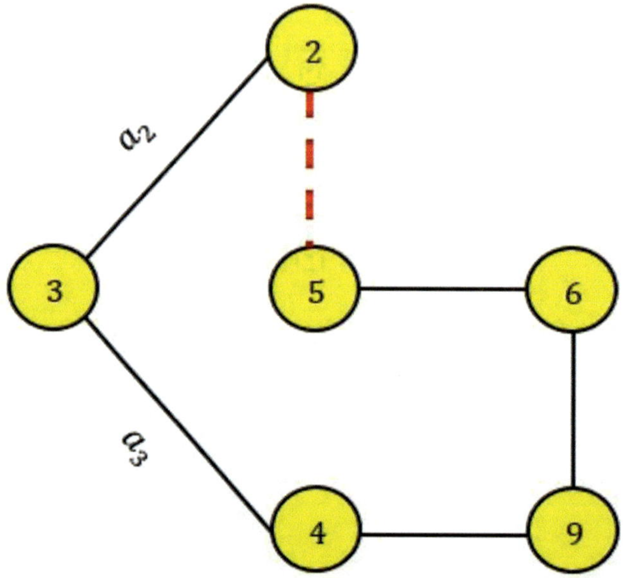

Figure 7.15.2 Fundamental circuit C_2.

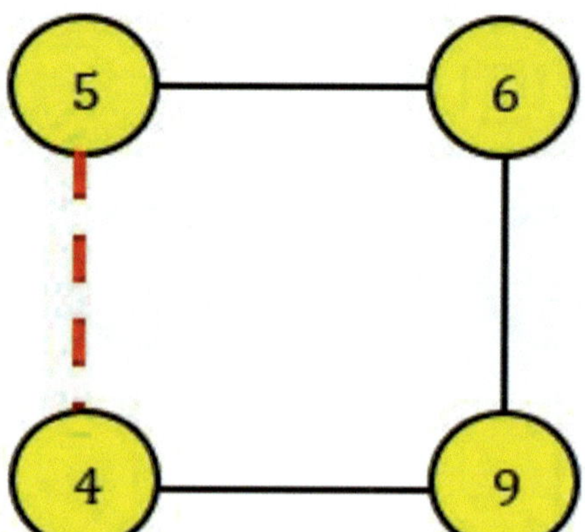

Figure 7.15.3 Fundamental circuit C_3

The path containing the failed links is comprised of the links $1 \to 2 \to 3 \to 4 \to 9 \to 11$. This path has three links in the failed state, which are: (1,2), (2,3) (3,4). We need an alternative path which is free of these three failed links.

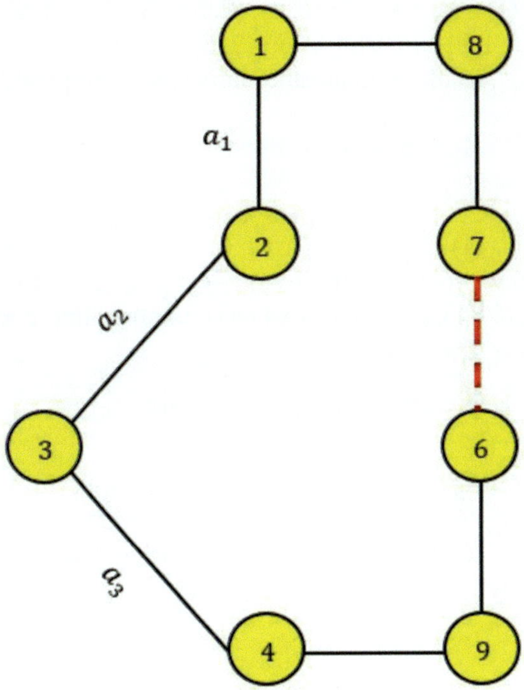

Figure 7.15.4 Fundamental circuit C_4.

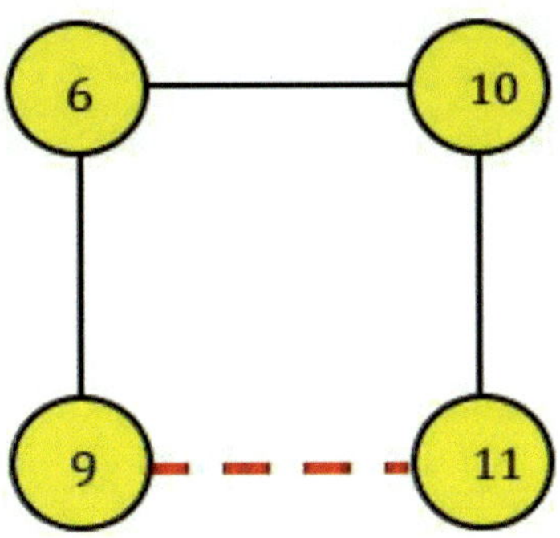

Figure 7.15.5 Fundamental circuit C_5.

The path free of the failed state will be comprised of the ring sums of the path '*P*' and those fundamental circuits which contain the failed links, so that the ring sum will be free of the failed links. These combinations are given below:

Possibility 1: $[P \bigoplus C_4] = [\{(1,2),(2,3),(3,4),(4,9),(9,11)\} \bigoplus \{(1,2),(2,3),(3,4),(4,9),(6,9),(6,7),(7,8),(1,8)\}] = [\{(1,8),(8,7),(7,6),(6,9),(9,11)\}]$, resulting in the alternative path comprised of five links is given by:

$$1 \rightarrow 8 \rightarrow 7 \rightarrow 6 \rightarrow 9 \rightarrow 11$$

.

Possibility 2: $[P \bigoplus C_4 \bigoplus C_3] =$ [{(1,2), (2,3), (3,4), (4,9), (9,11)} $\bigoplus${(1,2), (2,3), (3,4), (4,9), (6,9), (6,7), (7,8), (1,8)} $\bigoplus${(4,5), (5,6), (4,9), (6,9)}] = [{((1,8), (8,7), (7,6), (9,11), (4,5), (5,6), (4,9)}], the resulting alternative path comprised of 7 links is given by:

$$1 \rightarrow 8 \rightarrow 7 \rightarrow 6 \rightarrow 5 \rightarrow 4 \rightarrow 9 \rightarrow 11.$$

Possibility 3: $[P \bigoplus C_4 \bigoplus C_5] =$ [{(1,2), (2,3), (3,4), (4,9), (9,11)} $\bigoplus$ {(1,2), (2,3), (3,4), (4,9), (6,9), (6,7), (7,8), (1,8)} $\bigoplus${(6,9), (6,10), (9,11), (10,11)}] = [{(1,8), (8,7), (7,6), (6,10), (10,11)}], the resulting alternative five links path, which is given by:

$$1 \rightarrow 8 \rightarrow 7 \rightarrow 6 \rightarrow 10 \rightarrow 11.$$

Possibility 4: $[P \bigoplus C_1 \bigoplus C_2] =$ [{(1,2), (2,3), (3,4), (4,9), (9,11)} $\bigoplus$ {(1,8), (1,2), (2,7), (7,8)} $\bigoplus$ {(2,3), (3,4), (4,9), (9,6), (6,5). (5,2)}] = [{(1,8), (8,7), (7,2), (2,5), (5,6), (6,9), (9,11)}], resulting in a seven-link path given below:

$$1 \rightarrow 8 \rightarrow 7 \rightarrow -2 \rightarrow 5 \rightarrow 6 \rightarrow 9 \rightarrow 11.$$

Possibility 5: $[P \oplus C_1 \oplus C_2 \oplus C_3] =$ [{(1,2), (2,3), (3,4), (4,9), (9,11)} $\bigoplus$ {(1,8), (1,2), (2,7), (7,8)} $\bigoplus$ {(2,3), (3,4), (4,9), (9,6), (6,5). (5,2)}$\bigoplus$ {(4,5), (5,6), (6,9), (9,4)}] = [{(1,8), (8,7), (7,2), (2,5), (5,4), (4,9), (9,11)}], resulting in a seven-link path as given below:

$$1 \rightarrow 8 \rightarrow 7 \rightarrow 2 \rightarrow 5 \rightarrow 4 \rightarrow 9 \rightarrow 11.$$

Possibility 6: $[P \oplus C_1 \oplus C_2 \oplus C_5] =$ [{(1,2), (2,3), (3,4), (4,9), (9,11)} $\bigoplus${(1,8), (1,2), (2,7), (7,8)} $\bigoplus$ {(2,3), (3,4), (4,9), (9,6), (6,5), (5,2)} $\bigoplus$ {(6,10), (10,11), (11,9), ((9,6)}] = [{(1,8), (8,7), (7,2), (2,5), (5,6), (6,10), (10,11)}] , which give rise to a seven-link path as follows:

$$1 \rightarrow 8 \rightarrow 7 \rightarrow 2 \rightarrow 5 \rightarrow 6 \rightarrow 10 \rightarrow 11.$$

In conclusion, we have identified 6 different paths which are free of the failed links. Several paths are comprised of 7-link paths, and a few are comprised of 5-link paths. Depending on the requirement, one can select any one path as an alternative path to complete the task.

7.7 Concluding Remarks

In this chapter, we have considered stochastic networks for analysis of some reliability-related problems. For the maximum reliability route, we developed a new reconstruction, and evaluated a given stochastic network for a maximum reliability and identified the associated maximum reliability path. We compared our new approach with an earlier attempt, that was based on use of the functional recurrence relations, based on the dynamic programming approach. We also analysed the terminal-pair reliability of a stochastic network and obtained alternative paths when a specific path is found to be in the down state. We obtained alternative paths in a stochastic network which are free of the forbidden links.

References

A.S. Acampora, and M.J. Karol, an overview of light weight packed network, IEEE Networks, pp29-49, 1989.

M.O. Ball, Computing network reliability, Operations Research, 27(4), pp 823-838, 1979.

M.O. Ball, and G. Nemhauser, Matroids and Reliability Problems, Mathematics of Operations Research, 4(2), pp132-143, 1979.

S.P. Bansal, and S. Kumar, Maximum reliability route subject to K improvements in a directed network, IEEE Transactions on Reliability, R-26, pp103-105, 1977.

R. Bellman, and S.E. Dreyfus, Applied dynamic programming, Princeton University Press, 1962.

L. Fratta, and U.G. Montanari, A Boolean Algebra method for computing the terminal pair reliability in communication networks, IEEE Transactions on Circuit Theory, CT-20, No 3, 1973.

E. Hansler, G.K. McAuliffe, and R. S. Wilkow, Exact calculations for Network Reliability, Networks, Vol. 4, 95-112, 1974.

M. Herzberg, and S.J. Bye, An Optimal spare capacity assignment model for survivable networks with hop limits, Proc. Globecom' 94 held in San Franscico, California, Nov 27-Dec 1, 1994.

Y.H. Kim, K.E. Case, and P.M. Ghare, A method for computing complex system reliability, IEEE Transactions on Reliability, R-21, 215-219, 1972.

R. Kalaba, On some communication network problems, Combinatorial Anal. Proc. Symp. Appl. Maths, 10, 261-80, 1960.

S. Kumar, Maximum reliability route subject to M improvements in a non-directed network, IEEE Transactions on Reliability R-28, pp127-178, 1979.

S. Kumar, Sequential evaluation of terminal pair reliability in a reliability network, Microelectron Reliability, Vol 22, No 2, pp151-162, 1982.

S. Kumar, Optimization of protean systems: A review, APORS'94, Fukuoka, Japan, (Eds.) M. Fushimi and K. Tone, World Scientific Publishers, pp139-146, 1995.

S. Kumar, and H. Arora, K shortest paths and paths through K specified edges in a protean network, ASOR Bulletin, Vol 14, pp3-8, 1995.

S. Kumar, H. Arora, C. Chiera, and G. Raviganesh, Paths for self-healing Protean networks, Bulletin of the Allahabad Mathematical Society, B.N. Prasad Birth Centenary Commemoration Volume, 14, pp79-94, 1999.

S. Kumar, E. Munapo, O. Ncube, C. Sigauke, and P. Nyamhure, A minimum weight labelling method for determination of the shortest route in a non-directed network, Int J System Assur and Eng Management, 2013 Online publication DOI10.1007/s13198-012 (6 Jan 2013)

S. Kumar, E. Munapo, M. Lesaona, P. Nyamgure, Identification and application of virtual directions in a non-directed network: A labelling method for determination of Maximum reliability and the route, Communications in Dependability and Quality Management: An International Journal, Vol 19, No. 1, pp85-95, 2016.

S. Kumar, E. Munapo, P. Nyamgure, and T. Tawanda, Mathematics of OR: Significance and Application of virtual directions in reducing computational complexity in network optimization, Chapter 3 in Emerging Trends in Applied Research, Vol 3, Ed IS Chauhan, Integrated Publication, ISBN 978-93-93502-56-8, DOI: http://doi.org/10.22271/int.book.186, 2022

P.M. Lin, B.J. Leon, and T.C. Huang, A new algorithm for Symbolic Reliability analysis, IEEE Transactions on Reliability, R-25, pp2-16, 1976.

E. Munapo, C.E. Jones, and S. Kumar, A minimum incoming weight labelling method and its application to CPM network, ORiON, Vol 24, 2008, South African Journal of Operations Research.

N.J. Otway, Determination of the most reliable route in a reliability network, Proceedings of 2nd National Conference the Australian Society for Operations Research, Sydney pp627-647, 1975.

R. Patrovic, and S. Jovanovic, Two algorithms for determining the most reliable path of a network, IEEE Transactions on Reliability, R-28(2), pp115-119, 1979.

S.S. Srivastava, R.C. Garg, A.K. Govil, and S. Kumar, Optimal reliability route in a communication network, Microelectronics and Reliability, 11, 147-150, 1972.

8

Reconstruction Approach to Reliability and Maximum Flows in Networks

Abstract

In this chapter, we have discussed reliability and flow problems by reconstruction approaches. Reconstruction in each case is different; however, for the reliability problem, the purpose of reconstruction is to transform the problem to a manageable form where some existing tools can be used to find the reliability. For the network flows, the reconstruction aims to reduce complexity of the network and reach an equivalent network, where the maximal flows can be easily evaluated. Numerical illustrations have been given for both reconstruction approaches.

Keywords: Reliability problem, binary linear program, convex quadratic program, Hessian matrix, flows in networks, network reconstruction, maximum flows in a network.

8.1 Introduction

This chapter deals with two well-known areas, i.e., optimization in reliability and network flows. This discussion has been organised in two parts. Part A, deals with the reliability problem, which has been organized as Sections 8.2, 8.3, and 8.4. Part B deals with the network flows, which has been organized as Sections 8.5, 8.6, 8.7, and 8.8. Finally, some concluding remarks for both, i.e., the reliability and network flows, are presented in Section 8.9. Literature review has been provided in appropriate sections for both problems.

PART A: Reliability by Reconstruction Considerations

8.2 The Reliability Problem and Related Background

Reliability of a device is the probability that the device will function without failure under some specified conditions and for a given period. Accurate estimation of reliability is important for all situations, but it becomes crucial in situations where national security or human life is at stake. For example, situations such as aviation, health-related delivery systems, quality control, or for military operations, they all demand dependent systems; see Johnston et al. (2006). Devices used in these strategic situations must be reliable. Research on reliability evaluation of various devices is active and an ongoing field of research; see Li et al. (2017). There are several exact methods that can be used to evaluate the reliability of an equipment and many of these methods make use of mathematical models and their analysis. Standard methods such as linear integer models, dynamic programming models, and nonlinear integer models have been used in reliability context; see Bertsekas (2017). The weakness of dynamic approach is that complexity increases with an increase in the number of variables. Similarly, the complexity of integer models can easily become demanding. Many integer models are 'NP hard', and others are computationally demanding (see Arora and Barak 2009; Garey and Johnson 1979; Nemhauser and Wolsey 1988). Within these NP hard problems, nonlinear integer models are comparatively difficult and more demanding to solve than the linear ones. Exact approaches for linear and nonlinear integer models exist, but integer models are difficult to solve. Computational times for these exact methods can become excessive (Genova and Guliashki 2011). The reliability problem belongs to the nonlinear integer category. To alleviate these difficulties and keep the computational time manageable, heuristics that can quickly provide approximate estimations are desired in many situations. The heuristics for nonlinear optimization such as particle swarm optimization by Kennedy and Eberhart (1995), and genetic algorithms by Mitchell (1996) have proved to be effective for nonlinear integer models. Here in this chapter, we use the equal tendency idea based on Munapo (2019), which is discussed in Section 8.2.1.

8.2.1 Equal tendency theorem

The equal tendency theorem was used by Munapo (2007) to reformulate the reliability evaluation of a device. He used it for developing the bounds and for approximating reliability evaluation of a device. The equal tendency is the property that given various devices and their reliabilities, the product of the reliabilities tends to be maximal if the reliabilities of the various devices are approximately equal. This property is used to approximate the reliability of a system. Munapo (2016) also showed that the binary linear integer model can be solved in polynomial time by interior point algorithms (Gondzio 2012), which will be used in this chapter for reliability evaluation. First, we develop a nonlinear model for the typical reliability model, as discussed in Section 8.2.2.

8.2.2 Nonlinear form of a typical reliability model

Let y_{ij} be a binary variable and x_j be a nonnegative integer variable. Suppose that the total cost is restricted by a given quantity C_{Total}, and the objective is to choose a unit from each device such that the reliability product R_{Product} is maximal. The problem information is organized in the form of Table 8.1.

Table 8.1 The reliability problem.

	Device 1	**Device 2**	. . .	**Device n**
	$C_{1j}r_{1j}$	$C_{2j}r_{2j}$		$C_{nj}r_{nj}$
Parallel unit 1	$C_{11}r_{11}$	$C_{21}r_{22}$	. . .	$C_{n1}r_{n1}$
Parallel unit 2	$C_{21}r_{21}$	$C_{22}r_{22}$		$C_{n2}r_{n2}$
	$\vdots$	$\vdots$		$\vdots$
Parallel unit m	$C_{m1}r_{m1}$	$C_{m2}r_{m2}$		$C_{mn}r_{mn}$

The mathematical model will be given by relation (7.7) as given below.

$$
\begin{aligned}
&\text{Maximize } R_{\text{Product}} = \textstyle\prod_{j=1}^{j=n} x_j \\
&C_{11}y_{11} + C_{12}y_{12} + \cdots + C_{mn}y_{mn} \leq C_{\text{Total}} \\
&y_{11} + y_{12} + \cdots + y_{1m} = 1 \\
&y_{21} + y_{22} + \cdots + y_{2m} = 1 \\
&\vdots \\
&y_{n1} + y_{n2} + \cdots + y_{nm} = 1 \\
&x_1 = r_{11}y_{11} + r_{12}y_{12} + \cdots + r_{1m}y_{1m} \\
&x_2 = r_{21}y_{21} + r_{22}y_{22} + \cdots + r_{2m}y_{2m} \\
&\vdots \\
&x_n = r_{n1}y_{n1} + r_{n2}y_{n2} + \cdots + r_{nm}y_{nm}.
\end{aligned}
\tag{7.7}
$$

It may be noted that C_{ij}, r_{ij} are given and the unknown to be determined is x_j. The above is a nonlinear model, with integer variables x_j and y_{ij} as binary variables making it difficult to solve. However, this type of situation can be handled by using the equal tendency theorem. The features of the equal tendency theorem were noticed by Munapo (2007). The theorem states that the maximal reliability can be attained when devices selected have equal probability.

8.2.3 Evaluation of equal tendency theorem: Numerical illustration 8.1

We observe the equal tendency theorem by using a numerical illustration. Suppose we have 2 devices and 9 parallel units as given in Table 8.2. Also assume that the total costs cannot exceed 10 and that costs and reliabilities of the two devices are the same. The solution by exhaustive enumeration is given in Table 8.3.

Table 8.2 Numerical data of a trivial reliability problem.

	Device 1: C_{1j}, r_{1j}	**Device 2: C_{2j}, r_{2j}**
Parallel unit 1	$C_{11} = 1,\ r_{11} = 0.1$	$C_{21} = 1,\ r_{21} = 0.1$
Parallel unit 2	$C_{12} = 2,\ r_{12} = 0.2$	$C_{22} = 2,\ r_{22} = 0.2$
Parallel unit 3	$C_{13} = 3,\ r_{13} = 0.3$	$C_{23} = 3,\ r_{23} = 0.3$
Parallel unit 4	$C_{14} = 4,\ r_{14} = 0.4$	$C_{24} = 4,\ r_{24} = 0.4$
Parallel unit 5	$C_{15} = 5,\ r_{15} = 0.5$	$C_{25} = 5,\ r_{25} = 0.5$
Parallel unit 6	$C_{16} = 6,\ r_{16} = 0.6$	$C_{26} = 6,\ r_{26} = 0.6$
Parallel unit 7	$C_{17} = 7,\ r_{17} = 0.7$	$C_{27} = 7,\ r_{27} = 0.7$
Parallel unit 8	$C_{18} = 8,\ r_{18} = 0.8$	$C_{28} = 8,\ r_{28} = 0.8$
Parallel unit 9	$C_{19} = 9,\ r_{19} = 0.9$	$C_{29} = 9,\ r_{29} = 0.9$

For the problem in Table 8.2, an exhaustive search can be arranged as given in Table 8.3.

Table 8.3 Exhaustive search for the maximum reliability of two device systems.

Device 1 (x_1)	**Device 2(x_2)**	**Total cost**	**Product reliability**
0.1	0.9	10	0.09
0.2	0.8	10	0.16
0.3	0.7	10	0.21
0.4	0.6	10	0.24
0.5	0.5	10	0.25* the max
0.6	0.4	10	0.24
0.7	0.3	10	0.21
0.8	0.2	10	0.16
0.9	0.1	10	0.09

Note that the optimal solution obtained when the reliabilities of the two devices are equal, i.e., when $r_1 = 0.5$ and $r_2 = 0.5$ and the maximum product value is given by 0.25.

8.2.4 Evaluation of equal tendency: Numerical illustration 8.2

Similarly, let us examine one more trivial situation, and note that the number of combinations increase. Suppose that there are 3 devices and 7 parallel units and that the total cost cannot exceed 9; see Table 8.4. A detailed analysis is given in Table 8.5.

Table 8.4 Numerical data of a trivial reliability problem.

	Device 1: C_{1j}, r_{1j}	**Device 2: C_{2j}, r_{2j}**	**Device 3: C_{3j}, r_{3j}**
Parallel unit 1	$C_{11} = 1,\ r_{11} = 0.1$	$C_{21} = 1,\ r_{21} = 0.1$	$C_{31} = 1, r_{31} = 0.1$
Parallel unit 2	$C_{12} = 2,\ r_{12} = 0.2$	$C_{22} = 2,\ r_{22} = 0.2$	$C_{32} = 2, r_{32} = 0.2$
Parallel unit 3	$C_{13} = 3,\ r_{13} = 0.3$	$C_{23} = 3,\ r_{23} = 0.3$	$C_{33} = 3, r_{33} = 0.3$
Parallel unit 4	$C_{14} = 4,\ r_{14} = 0.4$	$C_{24} = 4,\ r_{24} = 0.4$	$C_{34} = 4, r_{34} = 0.4$
Parallel unit 5	$C_{15} = 5,\ r_{15} = 0.5$	$C_{25} = 5,\ r_{25} = 0.5$	$C_{35} = 5, r_{35} = 0.5$
Parallel unit 6	$C_{16} = 6,\ r_{16} = 0.6$	$C_{26} = 6,\ r_{26} = 0.6$	$C_{36} = 6, r_{36} = 0.6$
Parallel unit 7	$C_{17} = 7,\ r_{17} = 0.7$	$C_{27} = 7,\ r_{27} = 0.7$	$C_{37} = 7, r_{37} = 0.7$

Table 8.5 Exhaustive search for the maximum reliability of three device system.

Serial number	**Device 1 (x_1)**	**Device 2(x_2)**	**Device (x_3)**	**Total cost**	**Product reliability**
1(0.1)	0.1	0.7	0.1	9	0.007
2(0.1)	0.1	0.6	0.2	9	0.012
3(0.1)	0.1	0.5	0.3	9	0.015
4(0.1)	0.1	0.4	0.4	9	0.016
5(0.1)	0.1	0.3	0.5	9	0.015
6(0.1)	0.1	0.2	0.6	9	0.012
7(0.1)	0.1	0.1	0.7	9	0.007
1(0.2)	0.2	0.6	0.1	9	0.012
2(0.2)	0.2	0.5	0.2	9	0.020
...	...	...	...	9	...
6(0.2)	0.2	0.1	0.6	9	0.012
1(0.3)	0.3	0.5	0.1	9	0.015
2(0.3)	0.3	0.4	0.2	9	0.024
3(0.3)	0.3	0.3	0.3	9	0.027* max
...	...	...	...	...	...
5(0.3)	0.3	0.1	0.5	9	0.015
...	...	...	...	9	...
1(0.7)	0.7	0.1	0.1	9	0.007

Note that in the above table, there are 7 entries with (0.1), 6 entries with (0.2), 5 entries with (0.3), 4 entries with (0.4), 3 entries with (0.5), 2 entries with (0.6), and 1 entry with (0.7) and the maximum will be when all devices have equal probability of 0.3.

8.2.5 Evaluation of equal tendency: Numerical illustration 8.3

If the number of devices is increased to 4, then for the maximum reliability the total cost cannot exceed 8, as given in Table 8.6.

Table 8.6 Numerical data of a trivial reliability problem.

	Device 1: C_{1j}, r_{1j}	**Device 2**: C_{2j}, r_{2j}	**Device 3:** C_{3j}, r_{3j}	**Device 4**:
Parallel unit 1	$C_{11}= 1$, $r_{11} = 0.1$	$C_{21} = 1$, $r_{21} = 0.1$	$C_{31} = 1$, $r_{31} = 0.1$	$C_{41} = 1$ $r_{41} = 0.1$
Parallel unit 2	$C_{12} = 2$, $r_{12} = 0.2$	$C_{22} = 2$, $r_{22} = 0.2$	$C_{32} = 2$, $r_{32} = 0.2$	$C_{42} = 2$ $r_{42} = 0.2$
Parallel unit 3	$C_{13} = 3$, $r_{13} = 0.3$	$C_{23} = 3$, $r_{23} = 0.3$	$C_{33} = 3$, $r_{33} = 0.3$	$C_{43} = 3$ $r_{43} = 0.3$
Parallel unit 4	$C_{14} = 4$, $r_{14} = 0.4$	$C_{24} = 4$, $r_{24} = 0.4$	$C_{34} = 4$, $r_{34} = 0.4$	$C_{44} = 4$ $r_{44} = 0.4$
Parallel unit 5	$C_{15} = 5$, $r_{15} = 0.5$	$C_{25} = 5$, $r_{25} = 0.5$	$C_{35} = 5$, $r_{35} = 0.5$	$C_{45} = 5$ $r_{45} = 0.5$

Note that the number of entries can be evaluated for an exhaustive search, but the point is that exhaustive search will increase and will soon become unmanageable. The maximum will be when each device in Table 8.6 has equal reliability, when $r_{12} = r_{22} = r_{32} = r_{42} = 0.2$ and the maximum reliability will be 0.0016. These illustrations are not a coincidence; this is the outcome of the equal tendency theorem.

8.2.6 Proof of equal tendency theorem

The resource constraint in eqn (8.2a) implies that the selected devices cannot exceed available funds. Similarly, the sum over x_j is also restricted as given by eqn (8.2b).

$$C_{11}y_{11} + C_{12}y_{12} + \cdots + C_{nm}y_{nm} \leq C_{\text{Total}} \tag{8.2a}$$

$$x_1 + x_2 + \cdots + x_n \leq R_{\text{Total}} \tag{8.2b}$$

All the above illustrations can be generalized as eqn (8.3).

$$\text{Maximize } R_{\text{Product}} = \prod_{j=1}^{n} x_j \tag{8.3}$$

such that $x_1 + x_2 + \cdots + x_n \leq R_{\text{Total}}$.

The Lagrangian function (*L*) is given in eqn (8.4).

$$L = x_1 x_2 \ldots x_n - \mu(x_1 + x_2 + \cdots + x_n - R_{\text{Total}}), \tag{8.4}$$

where $\mu \geq 0$.

The Karush–Kuhn–Tucker (Jensen and Bard 2003; Taha 2017) conditions are given in eqn (8.5).

$$\frac{\partial L}{\partial x_j} = \frac{\partial L}{\partial \mu} = 0. \tag{8.5}$$

Values of these partial derivatives from eqn (8.5) are given by eqn (8.6)

$$\begin{aligned} x_2 x_3 \ldots x_n - \mu &= 0 \\ x_1 x_3 \ldots x_n - \mu &= 0 \\ &. \\ &. \\ x_2 x_3 \ldots x_{n-1} - \mu &= 0 \\ x_1 + x_2 + \cdots + x_n &= R_{\text{Total}} \ . \end{aligned} \tag{8.6}$$

Relation (8.6) simplify to:

$$\begin{aligned} x_2 x_3 \ldots x_n &= \mu \\ x_1 x_3 \ldots x_n &= \mu \\ &. \\ &. \\ x_2 x_3 \ldots x_{n-1} &= \mu \\ x_1 + x_2 + \cdots + x_n &= R_{\text{Total}} \ , \end{aligned} \tag{8.7}$$

which means that:

$$x_2 x_3 \ldots x_n = x_1 x_3 \ldots x_n = x_2 x_3 \ldots x_{n-1} = \mu. \tag{8.8}$$

The solution is given by eqn (8.9):

$$x_1 = x_2 = \cdots = x_n = \frac{R_{\text{Total}}}{n}. \tag{8.9}$$

8.3 A Linear Binary Model

Here we discuss a linear binary model, which will be used later for solving the reliability problem. Suppose the linear binary model is given by eqn (8.10).

$$\text{Maximize } c_1 z_1 + c_2 z_2 + \cdots + c_n z_n$$

such that

$$\begin{aligned} a_{11}z_1 + a_{12}z_2 + \cdots + a_{1n}z_n &= b_1 \\ a_{21}z_1 + a_{22}z_2 + \cdots + a_{2n}z_n &= b_2 \\ a_{m1}z_1 + a_{m2}z_2 + \cdots + a_{mn}z_n &= b_m, \end{aligned} \tag{8.10}$$

where $c_j \geq 0$, a_{ij}, and b_i are constants and z_j are binary variables, $i = 1, 2, \ldots, m$ and $j = 1, 2, \ldots, n$.

8.3.1 Special properties of binary numbers and their application

The linear binary model (8.10) can be reconstructed to a quadratic convex form by using an interesting property discussed by Munapo (2016). The special property is given in eqn (8.11).

$$x_j^2 + s_j^2 = 1, \tag{8.11}$$

where x_j is the binary variable, and s_j is a slack and when x_j is 1, then s_j is 0 and vice versa. The special feature of relation (8.11) is that it will force variables to assume binary values and it can be incorporated into a quadratic function as a part of the objective function. The linear binary problem in convex quadratic form is given in eqn (8.12).

Maximize $c_1x_1+c_2x_2+\cdots+c_nx_n+h(x_1^2+x_2^2+\cdots+x_n^2+s_1^2+\cdots+s_n^2)$ such that

$$\begin{aligned} a_{11}z_1 + a_{12}z_2 + \cdots + a_{1n}z_n &= b_1 \\ a_{21}z_1 + a_{22}z_2 + \cdots + a_{2n}z_n &= b_2 \\ &\cdots \\ a_{m1}z_1 + a_{m2}z_2 + \cdots + a_{mn}z_n &= b_m \\ x_1 + s_1 &= 1 \\ x_2 + s_2 &= 1 \\ &\cdots \\ x_n + s_n &= 1, \end{aligned} \tag{8.12}$$

where $h > 0$ is a large constant relative to the values of c_j.

Munapo (2016) showed that the convex quadratic objective in eqn (8.12) can be converted to linear form as given in eqn (8.13).

$$\begin{aligned} & c_1x_1 + c_2x_2 + \cdots + c_nx_n + h\left(x_1^2 + x_2^2 + \cdots + x_n^2 + s_1^2 + \cdots + s_n^2\right) \\ & = c_1x_1 + c_2x_2 + \cdots + c_nx_n + h\left(\left(x_1^2 + s_1^2\right) + \cdots + \left(x_n^2 + s_n^2\right)\right) \\ & = c_1x_1 + c_2x_2 + \cdots + c_nx_n + h\left((1) + (1) + \cdots + (1)\right) \\ & = c_1x_1 + c_2x_2 + \cdots + c_nx_n + nh \\ & = c_1x_1 + c_2x_2 + \cdots + c_nx_n + \text{constant}. \end{aligned} \tag{8.13}$$

8.3.2 Positive definiteness of the quadratic objective function

Let

$$F(X) = c_1x_1 + c_2x_2 + \cdots + c_nx_n + h\left(x_1^2 + x_2^2 + \cdots + x_n^2 + s_1^2 + \cdots + s_n^2\right). \tag{8.14}$$

Note that the function $F(X)$ has continuous second-order partial derivatives, and the Hessian matrix *H* exists and is given by eqn (8.15).

$$H = \begin{bmatrix} 2h & 0 & \cdots & 0 \\ 0 & 2h & \cdots & 0 \\ \cdots & \cdots & \cdots & \cdots \\ 0 & 0 & \cdots & 2h \end{bmatrix}. \tag{8.15}$$

The matrix *H* is positive definite since $X^T HX > 0 \; \forall X \neq 0$. More discussion on positive definiteness is given by Jensen and Bard (2003).

8.3.3 Binary model converted to convex quadratic form: Numerical illustration 8.4

Consider a linear binary problem, which we solve by converting into a convex quadratic form.

$$\text{Maximize } 10x_1 + 4x_2 + 19x_3$$

such that:

$$\begin{gathered} 2x_1 + 3x_2 + 4x_3 \leq 6, \\ 16x_1 + 5x_2 + 17x_3 \geq 21 \\ \vdots \\ 6x_1 + 17x_2 + 19x_3 \leq 23 \\ x_1 + s_1 = 1 \end{gathered}$$

$$x_2 + s_2 = 1$$

$$x_3 + s_3 = 1. \quad (8.16)$$

Here $s_1,\ s_2,\ s_3 \geq 0$ and binary. Let the large constant in this case be 1000. Solution of this problem came out to be: $x_1 = x_2 = 1,\ x_3 = 0$.

This approach will be used later for solving the reliability problem.

8.4 Practical Reliability Problems

Real-life reliability problems lack uniform pattern as has been considered in earlier numerical illustrations. Also, the total cost of the solution may not necessarily add up to the available resources. Since problem (7.7) in its present form is difficult to solve, we use the linear programming (LP) approach and control variables indirectly, which is discussed below.

8.4.1 Reliability problem as a linear integer model

We introduce bounds $(l_1,\ l_2)$ for the variable (x_j) and a slack variable for the resource constraint as given in eqn (8.17).

$$l_1 \leq x_j \leq l_2$$

$$C_{11}y_{11} + C_{12}y_{12} + \cdots + C_{nm}y_{nm} + s = C_{\text{Total}}. \quad (8.17)$$

From the equal tendency theorem, the solution is maximal if variables are the same and if all available resources are used. This is not enough for the solution to be maximal; x_j must be near l_2. This requirement of being near the l_2 can be approximated by the maximal of *R*. However, in this case, *R* is given by eqn (8.18).

$$R = x_1 + x_2 + \cdots + x_n. \quad (8.18)$$

In other words, there is a need to come up with a linear objective function of the form given in eqn (8.19).

$$\text{Minimize } \omega_1 (l_2 - l_1) + \omega_2 s - \omega_3 R, \quad (8.19)$$

where ω_1 and ω_2 are weights that can be selected. The weight ω_2 must be bigger than ω_1 to ensure that most of the resources are used. The linear objective function can be simplified to eqn (8.20).

$$\text{Minimize } \omega_1 l + \omega_2 s - \omega_3 R, \quad (8.20)$$

where,

$$l = l_2 - l_1. \tag{8.21}$$

Finally, the reliability problem (7.7) can be approximated as eqn (8.22).
Minimize $\omega_1 l + \omega_2 s - \omega_3 R$
such that,

$$\begin{aligned}
C_{11}y_{11} + C_{12}y_{12} + \cdots + C_{mn}y_{mn} &\leq C_{\text{Total}} \\
x_1 = r_{11}y_{11} + r_{12}y_{12} + \cdots &+ r_{1m}y_{1m} \\
x_2 = r_{21}y_{21} + r_{22}y_{22} + \cdots &+ r_{2m}y_{2m} \\
&\vdots \\
x_n = r_{n1}y_{n1} + r_{n2}y_{n2} + \cdots &+ r_{nm}y_{nm} \\
y_{11} + y_{12} + \cdots + y_{1m} &= 1 \\
y_{21} + y_{22} + \cdots + y_{2m} &= 1 \\
&\vdots \\
y_{n1} + y_{n2} + \cdots + y_{nm} &= 1 \\
R = x_1 + x_2 + \cdots &+ x_n \\
l_1 \leq x_1 \leq l_2, \ldots, l_1 \leq x_n \leq l_2 \text{ and } l &= l_2 - l_1.
\end{aligned} \tag{8.22}$$

8.4.2 Numerical illustration 8.5

Consider the reliability problem given in Tables 8.7 and 8.8.

Table 8.7 Problem data.

	Device 1	Device 2	Device 3	Device 4
	c_{1j}, r_{1j}	c_{2j}, r_{2j}	c_{3j}, r_{3j}	c_{4j}, r_{4j}
Parallel unit 1	20, 0.1	30, 0.8	10, 0.2	15, 0.3
Parallel unit 2	30, 0.2	40, 0.16	20, 0.3	20, 0.4
Parallel unit 3	40, 0.3	50, 0.24	30, 0.4	25, 0.5
Parallel unit 4	50, 0.4	60, 0.32	40, 0.5	30, 0.6
Parallel unit 5	60, 0.5	70, 0.40	50, 0.6	40, 0.7

The LP model for the above problem is given by eqn (8.23):
Minimize $10l + 100s - 10R$
such that,

$$20y_{11} + 30y_{12} + \cdots + 40y_{mn} \leq 120$$

$$
\begin{aligned}
x_1 &= 0.1y_{11} + 0.2y_{12} + 0.3y_{13} + 0.4y_{14} + 0.5y_{15} \\
x_2 &= 0.8y_{21} + 0.16y_{22} + 0.24y_{23} + 0.32y_{24} + 0.40y_{25} \\
x_3 &= 0.2y_{31} + 0.3y_{32} + 0.4y_{33} + 0.5y_{34} + 0.6y_{35} \\
x_4 &= 0.3y_{41} + 0.4y_{42} + 0.5y_{43} + 0.6y_{44} + 0.7y_{45} \\
& y_{11} + y_{12} + y_{13} + y_{14} + y_{15} = 1 \\
& y_{21} + y_{22} + y_{23} + y_{24} + y_{25} = 1 \\
& y_{31} + y_{32} + y_{33} + y_{34} + y_{35} = 1 \\
& y_{41} + y_{42} + y_{43} + y_{44} + y_{45} = 1 \\
& R = x_1 + x_2 + x_3 + x_4 \\
& l_1 \le x_1 \le l_2, l_1 \le x_2 \le l_2, l_1 \le x_3 \le l_2, l_1 \le x_4 \le l_2 \\
& \text{and } l = l_2 - l_1.
\end{aligned} \tag{8.23}
$$

Note that $y_{11}, y_{12}, \ldots, y_{45}$ are binary variables. The solution comes out to be:

$$y_{13} = y_{23} = y_{31} = y_{42} = 1 \text{ and } x_1 = 0.3,\ x_2 = 0.24,\ x_3 = 0.2,\ x_4 = 0.4$$

$$R_{\text{Product}} = x_1 x_2 x_3 x_4 = 0.3 \times 0.24 \times 0.2 \times 0.4 = 0.00576.$$

This solution is shown in bold in Table 8.8.

Table 8.8 Solution to problem in Table 8.7.

	Device 1	**Device 2**	**Device 3**	**Device 4**
	c_{1j}, r_{1j}	c_{2j}, r_{2j}	c_{3j}, r_{3j}	c_{4j}, r_{4j}
Parallel unit 1	20, 0.1	30, 0.8	**10, 0.2**	15, 0.3
Parallel unit 2	30, 0.2	40, 0.16	20, 0.3	**20, 0.4**
Parallel unit 3	**40, 0.3**	**50, 0.24**	30, 0.4	25, 0.5
Parallel unit 4	50, 0.4	60, 0.32	40, 0.5	30, 0.6
Parallel unit 5	60, 0.5	70, 0.40	50, 0.6	40, 0.7

8.4.3 Strength of the proposed approach

The reliability problem in the form (8.1) is very difficult to solve. In fact, it is NP hard, and we are not aware of any algorithm that can solve such a problem in polynomial time (Fortnow 2013). Here, we are attempting to make things easier by approximating to a linear form. The linear form, which is the general binary problem, can be solved in polynomial time by converting to convex quadratic form, as shown by Munapo (2016). Any linear binary problem can be transformed into convex quadratic problem. Gondzio (2012) showed that interior point algorithms can solve convex quadratic problem in polynomial time.

PART B: Flows in Networks

8.5 Introduction, Literature Review to Network Flows

Maximum flow problem is a network-based problem where the objective is to determine the maximum flow of units that can be moved from the source node to the sink node in some time interval. Network edges may represent pipes, electricity cables, and roads, while the nodes represent tanks, dams pipe junctions, road junctions, sub-power stations, and cities. Flow in the network edges may be water, oil, sewage, electricity, and vehicles, among others. Several algorithms have been developed to solve the maximum flow problem; see Kyi & Naing (2018), Munapo et al. (2021), Akram et al. (2022), Munapo et al. (2022), and Tawanda et al. (2023a, 2023b, 2024). Applications of the maximum flow problem include but not limited to:

(1) Network optimization: Maximizing network flow is crucial in telecommunications, transportation, and logistics to optimize network performance and efficiency.
(2) Scheduling: Maximum flow algorithms are used in scheduling tasks, jobs, and resources to maximize productivity and minimize delays.
(3) Resource allocation: The maximum flow problem is applied in resource allocation, such as allocating processing power, memory, and bandwidth in computer systems.
(4) Supply chain management: Maximum flow algorithms are used to optimize supply chain networks, managing inventory, and shipping routes.
(5) Traffic flow: The maximum flow problem is applied in traffic flow management to optimize traffic light timing and reduce congestion.
(6) Hydraulic networks: Maximum flow algorithms are used to optimize water supply networks, managing water flow and pressure (Tawanda et al. 2023b).
(7) Electrical networks: The maximum flow problem is applied in electrical networks to optimize power flow and minimize losses.
(8) Data transfer: Maximum flow algorithms are used to optimize data transfer in computer networks, maximizing throughput, and minimizing latency.

(9) Manufacturing systems: The maximum flow problem is applied in manufacturing systems to optimize production flow and minimize bottlenecks.
(10) Healthcare: Maximum flow algorithms are used in healthcare to optimize patient flow, managing hospital resources, and minimizing wait times.

8.6 Network Flows by the Reconstruction Algorithm

Given below is a simple idea of the network reconstruction algorithm developed by Munapo et al. (2022). The algorithm is simple, and it reduces the network size as the number of iterations increases. Unlike other existing algorithms that make use of augmenting path technique to compute the maximum flow, this algorithm reconstructs a reduced network that maintains the maximum flow value of the original network. The algorithm reduces any given network to a linear network or a simple network where the maximum flow can be easily determined.

Network reconstruction algorithm

Input: Directed flow network

Output: Maximum flow value

1. Reduce the network measures to their simplest values
2. Identify an outmost arc or path in the network. Remove this identified arc or path in such a way that the reduced (reconstructed) network has the same maximum flow value as the original network.
3. Using the reduced network, go to Step 2 until the network is small and that the maximum flow is easy and obvious to determine.

8.7 Numerical Illustration 8.6

Given in Figure 8.1 is a directed maximum flow network (Srinivasan 2017) with nodes 1 and 7 as the source and sink, respectively. The arc weights denote the capacities. The aim is to determine the maximum flow value from the source node 1 to the sink node 7.

Solution

Iteration 1

Simultaneously removing arcs (3, 6) and (2, 5), the respective bottleneck capacities of 20 and 10 units can be added to paths $(3 \rightarrow 4 \rightarrow 6)$ and $(2 \rightarrow 4 \rightarrow 5)$, respectively, without changing the maximum flow value as shown in Figure 8.2.

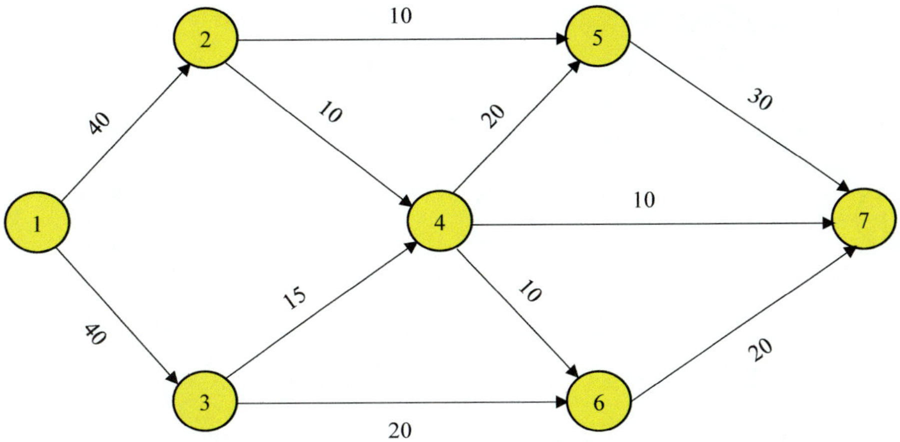

Figure 8.1 Seven nodes maximum flow problem.

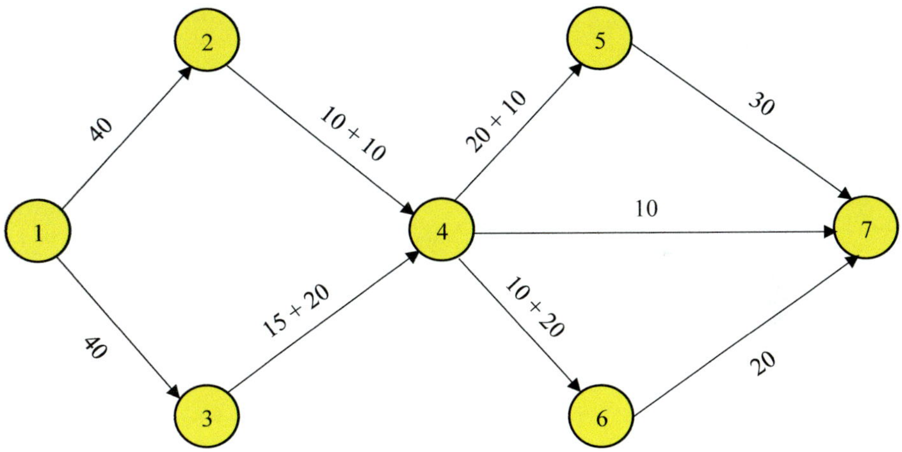

Figure 8.2 Reconstructed network after iteration 1.

Iteration 2

Removing arc (4, 7) with bottleneck capacity 10 units. Adding this capacity to path $(4 \rightarrow 6 \rightarrow 7)$. The resultant reconstructed network is given in Figure 8.3.

Iteration 3

Simultaneously removing paths $(1 \rightarrow 3 \rightarrow 4)$ and $(4 \rightarrow 5 \rightarrow 7)$, the respective bottleneck capacities of 35 and 30 units can be added to paths $(1 \rightarrow 2 \rightarrow 4)$ and $(4 \rightarrow 6 \rightarrow 7)$, respectively, without changing the maximum flow value as shown in Figure 8.4.

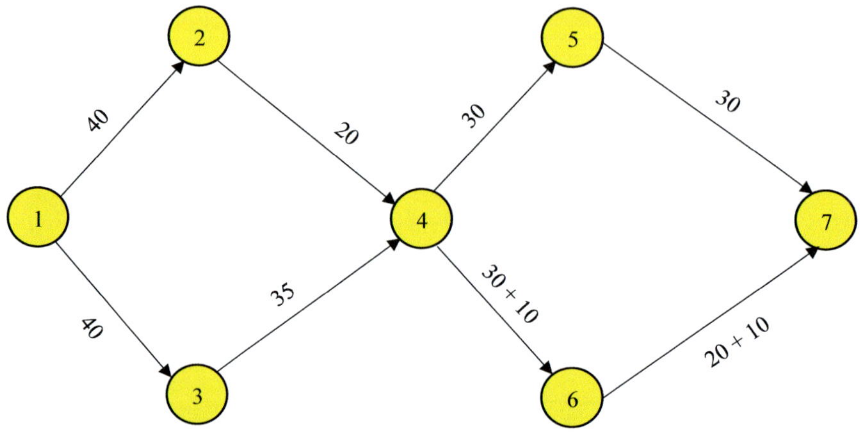

Figure 8.3 Reconstructed network after iteration 2.

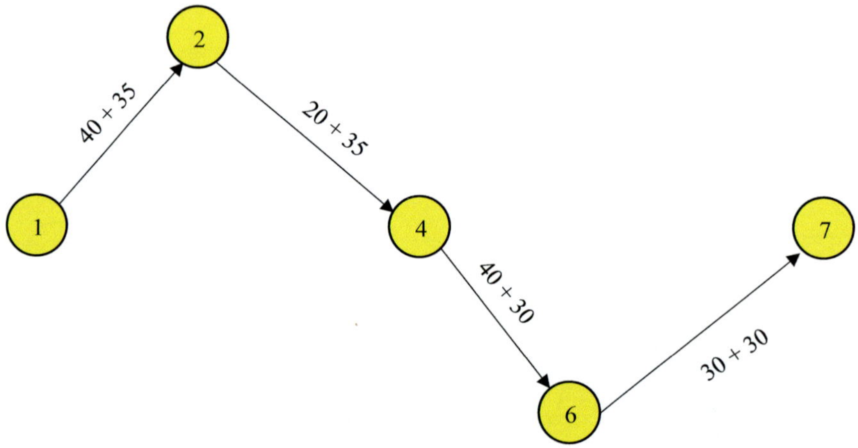

Figure 8.4 Reconstructed network after iteration 3.

The maximum flow value is obviously 55 units. This amount (Figure 8.4) is the bottleneck capacity of path $(1 \to 2 \to 4 \to 6 \to 7)$. This solution is the same as the one computed in Srinivasan (2017) using the flow augmenting path procedure.

8.7.1 Numerical illustration 8.7

Consider a 9-nodes network in Kinsley and Maheswari (2016), where nodes S and T are the source and sink, respectively. All other nodes are the intermediate nodes. The weights on the arcs represent the flow rates as shown in Figure 8.5.

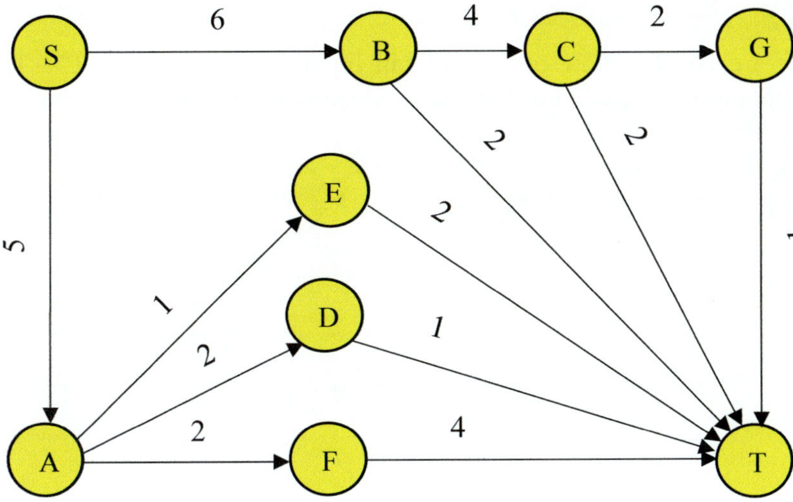

Figure 8.5 Nine-node maximum flow problem.

Solution

Iteration 1

Removing paths (A→E→T) and (A→D→T). Respective bottleneck capacities can be added to path (A→F→T) without changing the maximum flow value as shown in Figure 8.6.

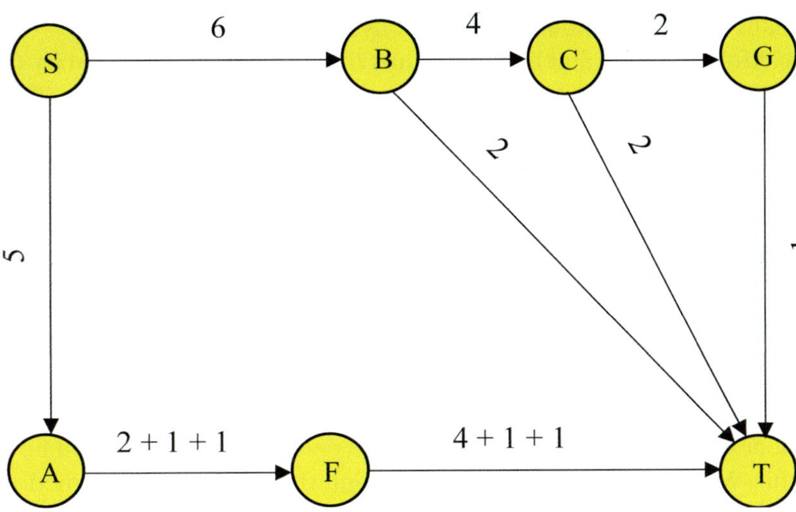

Figure 8.6 Reconstructed network after iteration 1.

Iteration 2

Removing arc (*B*, *T*). Flow of 2 units can be added to path (B→C→T) without changing the maximum flow value as shown in Figure 8.7.

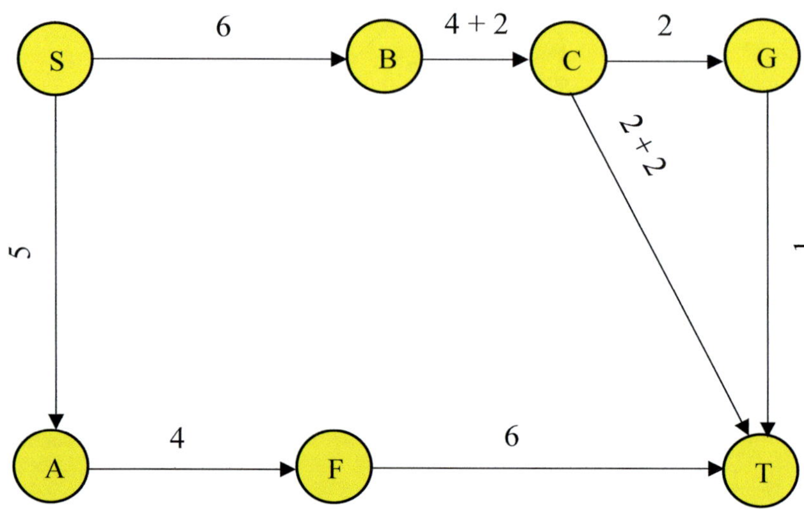

Figure 8.7 Reconstructed network after iteration 2.

Iteration 3

Removing arc (*C*, *T*). Flow of 4 units can be added to path (C→G→T) without changing the maximum flow value as shown in Figure 8.8.

The maximum flow value is 9 units, thus the total bottleneck capacity (5 + 4) of paths ($S \rightarrow B \rightarrow C \rightarrow G \rightarrow T$) and ($S \rightarrow A \rightarrow F \rightarrow T$), respectively. This is the same solution found by Kinsley and Maheswari (2016) and Tawanda et al. (2023a) using different algorithms.

8.8 Case study: Pyigyitagon Township Water Distribution Network (WDN), Mandalay, Myanmar

The satellite image showing the water distribution network in (yellow) for Pyigyitagon Township is shown in Figure 8.10. The respective graph representation of the WDN is also given in Figure 8.9. Nodes 1 and 6 are the source and sink nodes, respectively. The objective is to determine the maximum flow value in this WDN from node 1 to node 6 using the network reconstruction algorithm.

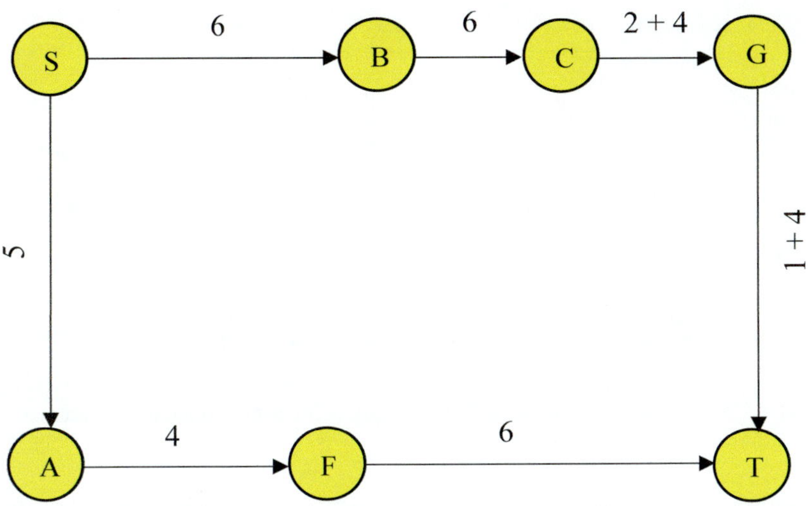

Figure 8.8 Reconstructed network after iteration 3.

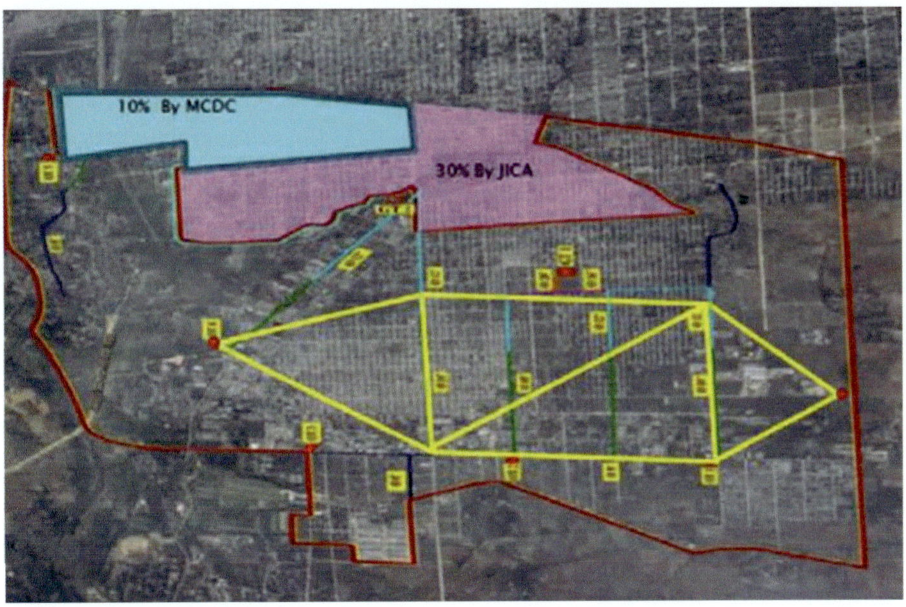

Figure 8.9 WDN in Pyigyitagon Township, Myanmar (Akram et al. 2022; Kyi & Naing 2018).

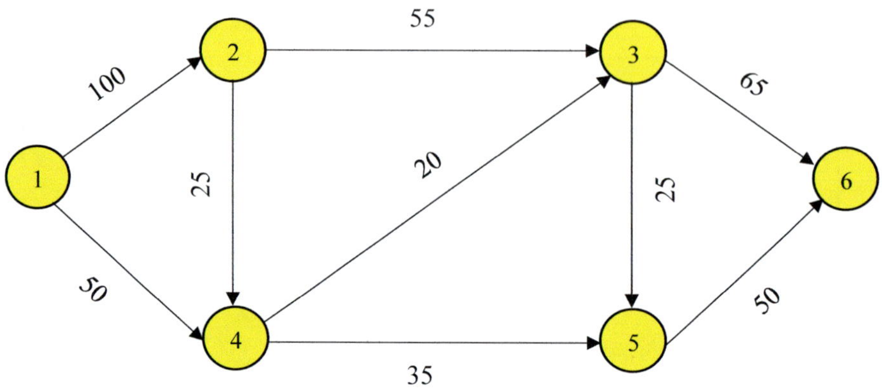

Figure 8.10 Water distribution network in Pyigyitagon Township, Mandalay, Myanmar.

Solution

In this section, network reconstruction approach is applied to determine the maximum flow in the network given in Figure 8.10.

Solution

Iteration 1

Removing arc (2, 4) with the capacity of 25 units. Subtracting this capacity from arc (1, 2) and adding it to arc (1, 4) give us a reconstructed network (Figure 8.11) with the same maximum flow value as the original network.

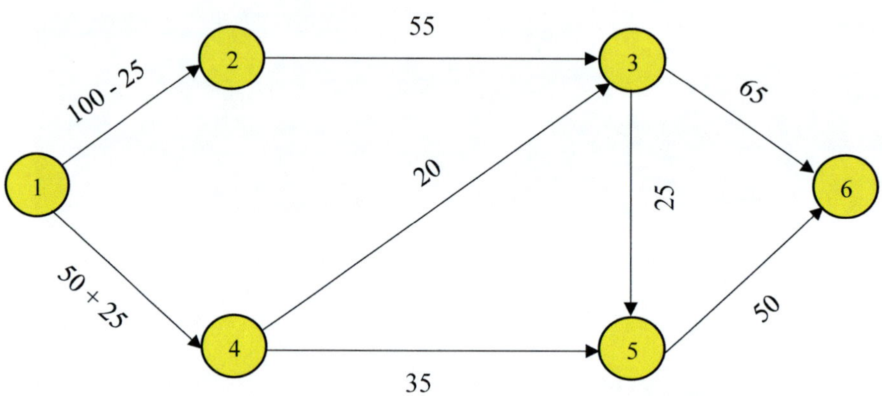

Figure 8.11 Reconstructed network after iteration 1.

Iteration 2

Removing path ($1 \rightarrow 2 \rightarrow 3$) with bottleneck capacity of 55 units and add this capacity to path ($1 \rightarrow 4 \rightarrow 3$) without changing maximum flow value as shown in Figure 8.12.

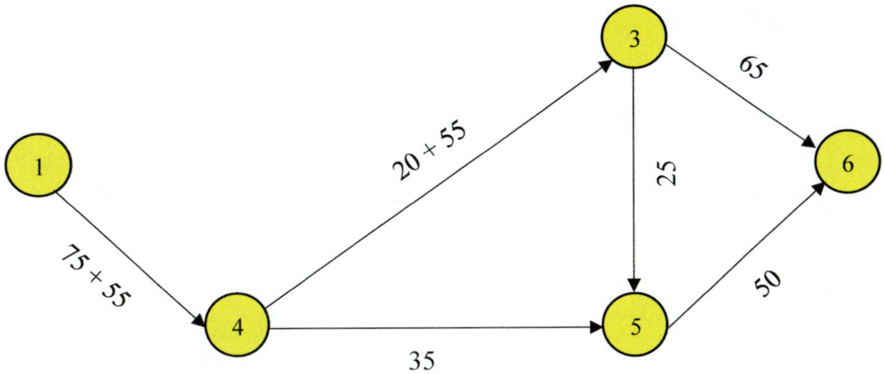

Figure 8.12 Reconstructed network after iteration 2.

Iteration 3

Removing arc (3, 6) with a capacity of 65 units and adding this capacity to path ($3 \rightarrow 5 \rightarrow 6$) without changing the maximum flow value as shown in Figure 8.13.

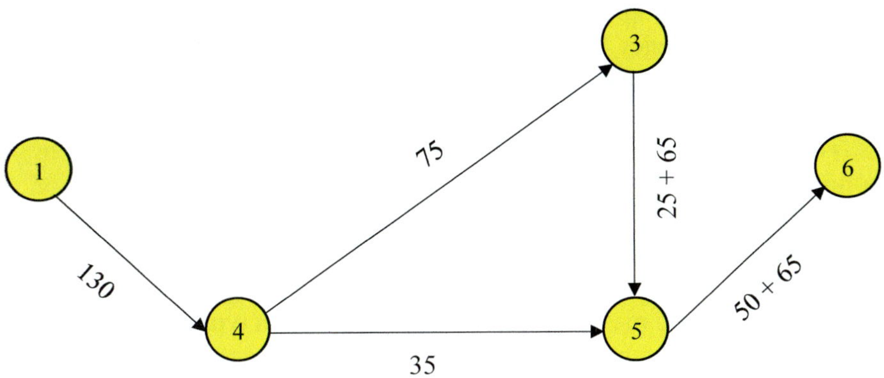

Figure 8.13 Reconstructed network after iteration 3.

Iteration 4

Removing arc (4, 5) with a capacity of 35 units and adding this capacity to path ($4 \rightarrow 3 \rightarrow 5$) without changing the maximum flow value as shown in Figure 8.14.

Since the reduced network cannot be reduced further, the maximum flow value is now obviously the bottleneck of the path ($1 \rightarrow 4 \rightarrow 3 \rightarrow 5 \rightarrow 6$), which is the minimum of $\{130, 110, 125, 115\} = 110$ Litres/sec. This solution is the same as the one computed in Kyi & Naing (2018) using the Ford–Fulkerson algorithm.

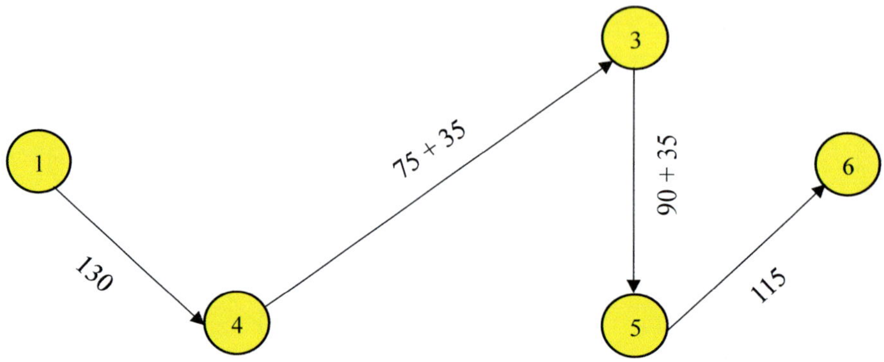

Figure 8.14 Reconstructed network after iteration 4.

8.9 Conclusion

In Part A of this chapter, we have presented several applications of the equal tendency algorithm to approximate the general reliability model. With the equal tendency algorithm, variables in the optimal solution tend to assume equal values. Linear programming and linear constraints are then used to enforce the equal tendency theorem to come up with a linear integer model, which is easier to solve than the original nonlinear form. The linear binary integer problem can be solved in polynomial time by interior point algorithms. The above discussion leads us to solve a nonlinear integer model in polynomial time is a notable achievement. The approximated linear form of the reliability model is a linear binary integer model, which, as shown by Munapo (2016), can be solved in polynomial time.

In Part B of this chapter, the network reconstruction algorithm has been applied to determine the maximum flow values in different types of networks. The algorithm does not make use of augmenting paths to determine the maximum flow. The algorithm reduces the size of the network by removing nodes and arcs while maintaining the maximum flow value unchanged of the original network. In general, the number of iterations required to compute the maximum flow are reasonable.

It is worth investigating the reconstruction approach, discussed in this chapter in some other situations.

References

M. Akram, A. Habib, T. Allahviranloo, A new maximal flow algorithm for solving optimization problems with linguistic capacities and flows,

Information Sciences 612, 201–230, 2022, https://doi.org/10.1016/j.ins.2022.08.068

S. Arora, B. Barak, Computational complexity: a modern approach, 1st Edition, Cambridge University Press, Cambridge, 2009.

D. P. Bertsekas, Dynamic programming and optimal control, vol 1, 4th Edition, Athena Scientific, Belmont, 2017.

F. Fortnow, The golden ticket: P, NP, and the search for the impossible, 1st edn. Princeton University Press, Princeton, 2013.

M. R. Garey, D. S. Johnson, Computers intractability: a guide to the theory of NP Completeness. W. H. Freeman, San Francisco, 1979.

K. Genova and V. Guliashki, Linear integer programming methods and approaches—a survey. Cybern Inf Technol 11(1):3–25, 2011.

J. Gondzio, Interior point methods 25 years later. Eur J Oper Res 218(2012):587–601, 2012. https://doi.org/10.1016/j.ejor.2011.09

P. A. Jensen, J. F. Bard, Operations research models and methods. Wiley, New York, 2003.

W. Johnston, J. L. Quigley, L. A. Walls, Optimal allocation of reliability tasks to mitigate faults during new product development. IMA J Manag Math 17(2):159–169, 2006.

J. Kennedy, R. C. Eberhart, Particle swarm optimization. In: Proceedings of IEEE international conference on neural networks, vol. IV, IEEE Service Centre, Piscataway, pp 1942–1948, 1995.

A. A. Kinsley and B. U. Maheswari, Design of algorithms to maximum flow problems in the flow networks, International Journal of Science, Engineering and Technology Research, Vol. 5, No. 4, pp. 1015–1018, 2016.

M. T. Kyi, & L. L. Naing, Application of Ford-Fulkerson algorithm to maximum flow in water distribution pipeline network. *International Journal of Scientific and Research Publications*, *8*(12), 306-310, 2018. http://dx.doi.org/10.29322/IJSRP.8.12.2018.p8441

P. Li, C. Li, W. Dang, Accelerated reliability demonstration testing design based on reliability allocation of environmental stresses. Qual Reliab Eng Int 33(7):1425–1435, 2017. https://doi.org/10.1002/qre.2115

M. Mitchell, An introduction to genetic algorithms. MIT Press, Cambridge, 1996

E. Munapo, African journal of science and technology (AJST), science and engineering series, vol 8, no 2, pp 33–43, 2007.

E. Munapo, Solving the binary linear programming model in polynomial time. Am J Oper Res 6:1–7, 2016. https://doi.org/10.4236/ajor.2016.61001

E. Munapo, The equal tendency theorem: A new heuristic for the reliability model, International Journal of System Assurance and Engineering Management, Vol. 10, No 5, pp 918-924, 2019, https://doi:.org/10.1007/s13198-019-00821-w

E. Munapo, S. Kumar, & K. Mpeta, A new direction to the maximal flow problem: a route merging approach. *Appl. Math*, *15*(4), 487-495, 2021.

E. Munapo, T. Tawanda, P. Nyamugure, & S. Kumar, Maximum flow by network reconstruction method. In *International Conference on Intelligent Computing & Optimization* (pp. 926-935), 2022, October. Cham: Springer International Publishing.

G. Nemhauser, L. A. Wolsey, Integer and combinatorial optimization, 1988, Wiley, New York.

G. Srinivasan, Operations research: Principles and applications. 3rd Edition, 2017, PHI Learning Pvt. Ltd.

H. A. Taha, Operations research: an introduction, 10th Edition, 2017. Pearson Educators, Bengaluru

T, Tawanda, P. Nyamugure, S. Kumar, & E. Munapo, An intelligent node labelling maximum flow algorithm. *International Journal of System Assurance Engineering and Management*, *14*(4), 1276-1284, 2023a.

T. Tawanda, P. Nyamugure, E. Munapo & S. Kumar, Extreme Min–Cut Max–Flow Algorithm. *International Journal of Applied Metaheuristic Computing (IJAMC)*, *14*(1), 1-16, 2023b.

T. Tawanda, S. Kumar, E. Munapo, & P. Nyamugure, Application of Network Reconstruction Algorithm to Compute Maximum Flow for Water Supply Network: A Case Study of the City of Bulawayo, Zimbabwe. Contemporary Mathematics, 5(2), 2173-2184, 2024.

Index

About the Authors

Elias Munapo holds a B.Sc. (Hons) in Applied Mathematics (1997), M.Sc. in Operations Research (2002) and a Ph.D. in Operations Research (2010), all from the National University of Science and Technology (NUST), Zimbabwe. He also has a certificate in outcomes-based assessment in Higher Education and Open Distance Learning, University of South Africa (UNISA) and a certificate in University Education Induction Program, University of KwaZulu-Natal (UKZN). He is a Professional Natural Scientist certified by the South African Council for Natural Scientific Professions (SACNASP), 2012; and NRF rated in South Africa. He has published/co-published over 120 articles and two books. He has also edited/co-edited 7 books and is a guest editor of Applied Sciences, Algorithms and Next Energy journals which are under MDPI. He has supervised/co-supervised eleven doctoral students and over 30 students at Master level. He is a member of the Operations Research Society of South Africa, Executive Committee Member 2012-13, South African Council for Natural Scientific Professions (SACNASP) as a Certified Natural Scientist, European Conference on Operational Research (EURO) and the International Federation of Operations Research Societies (IFORS), a member of the International Conference on Optimization (ICO) and is part of the team preparing to host ICO 2026 at Johannesburg.

Santosh Kumar OAM received his Ph.D. degree in Operations Research from Delhi University. He is author and co-author of over 210 papers and 4 books in the field of Operations Research. His contributions in the field of OR have been recognized in the form of the Ren Pots' award from the Australian Society for operations Research (ASOR) in 2009 and a recognition award from the South African OR Society as a non-member of the society and a non-resident of South Africa in 2011. He was the President of the Asia Pacific Operations Research societies (1995-97), where ASOR was a member along with 7 other countries in the region. He is currently an Honorary Professor at RMIT University, Melbourne. He is a Fellow of the Institute of Mathematics and its Applications, UK, a life member of the ASOR and a life Senior

Member of the Operations Research Society of India. On 14 June 2021, he was awarded a medal of the Order of Australia (OAM).

Philimon Nyamugure gained a B.Sc. (Hons) in 1998, an M.Sc. in 2002 and a Post Graduate Diploma in Higher Education in 2013, all from NUST, and a Ph.D. in Statistics, University of Limpopo, 2017. He joined NUST in 2003 and was Chairperson for the Departments of Applied Mathematics (2009-2013) and Chairman, Department of Statistics and Operations Research (2013-2019). He became Executive Dean of the Faculty of Applied Science from 2020 to date and a University Senator from 2009 and Councillor from 2020 to date. He received an award for the best Ph.D. presenter in the School of Mathematical and Computer Sciences, University of Limpopo. Best Senior presenter, Faculty of Applied Sciences, and in 2017 was awarded a Fulbright Research Award for African Scholars. He is involved in several community engagement programs including National University of Science and Technology Schools Enrichment Program (NUSTSEP). He is currently an Executive Dean and a Professor at NUST.

Trust Tawanda is a seasoned academic and researcher with a strong background in Operations Research and Statistics. He holds a Bachelor of Science (Hons) in Operations Research and Statistics (2013), a Master of Science in Operations Research and Statistics (2017), and a Ph.D. in Operations Research (2024) from the National University of Science and Technology (NUST) in Zimbabwe. Currently, he is pursuing a Post Graduate Diploma in Higher and Tertiary Education (PGDHTE) at Great Zimbabwe University (GZU). As a Senior Lecturer in the Department of Statistics and Operations Research at NUST's Faculty of Applied Sciences, he teaches and conducts research in his areas of expertise, which include network optimization problems, transportation problems, travelling salesman problems, and assignment problems. He has a notable publication record, with papers and book chapters to his credit. His academic journey and research endeavors demonstrate his commitment to advancing knowledge in Operations Research.

For Product Safety Concerns and Information please contact our
EU representative GPSR@taylorandfrancis.com Taylor & Francis
Verlag GmbH, Kaufingerstraße 24, 80331 München, Germany